Lucas Lima Gonçalves
Bruno T. de Farias
Victor A. N. Magalhães

Diesel Engine for Torque Gain with Mobile Compression Chamber

Lucas Lima Gonçalves
Bruno T. de Farias
Victor A. N. Magalhães

Diesel Engine for Torque Gain with Mobile Compression Chamber

Developed in CAD Tool

ScienciaScripts

Imprint

Cover image: www.ingimage.com

This book is a translation from the original published under ISBN 978-613-9-63040-0.

Publisher:
Sciencia Scripts
is a trademark of
Dodo Books Indian Ocean Ltd. and OmniScriptum S.R.L publishing group

120 High Road, East Finchley, London, N2 9ED, United Kingdom
Str. Armeneasca 28/1, office 1, Chisinau MD-2012, Republic of Moldova, Europe
Printed at: see last page
ISBN: 978-620-7-76486-0

SUMMARY

We dedicate this work to our families and friends, who have always supported us and, above all, to our teachers, who have given us the basis and sufficient support to make this happen.

ACKNOWLEDGMENTS

First of all, I would like to thank my family, who provided me with enough structure during my undergraduate years so that I could reach my final goal. To the teachers, technicians and staff at the Federal University of the Jequitinhonha and Mucuri Valleys, who have always dedicated themselves to providing the best education for us students. The city of Diamantina and the people I met here, who made this journey more enjoyable and helped me to complete this important stage in my life.

FARIAS, B. T. de, 2017

ACKNOWLEDGMENTS

I would first like to thank God. He is the foundation, responsible for me achieving this goal. To my parents, who gave me all the necessary support and unconditional support at all times, making it possible for me to get to where I am today. To my sisters, who were a second mother to me and helped me at important moments, making all the difference to my learning. To my friends, made in Diamantina and those from long ago, who were there for me in happy and not-so-happy moments when my family couldn't be there. To the masters and technicians, who have been a fundamental part in enabling me to have enough knowledge to do this work. Last but not least, I would like to give a special thanks to Professor Geraldo Darlan, who gave me my first notions of mechanics and thus encouraged this project, and to Prof. Dr. Victor Magalhaes, who believed in our idea and agreed to guide us in this work.

GONÇALVES, L. L, 2017

SUMMARY

With the growing quest to optimize internal combustion engines, more and more people are trying to obtain better performance without increasing fuel consumption. With these improvements in mind, the aim of this study is to develop a computer model that improves diesel cycle engines in order to obtain more torque. To do this, a change is proposed between engine times where compression is lengthened by another half turn of the crankshaft and at this point the combustion time occurs. This change is made possible by the installation of a movable compression piston, which will maintain compression in the chamber after the piston has compressed the mixture. When it reaches a midpoint, the compression will be increased so that the air-fuel mixture combusts to extract a higher torque. The entire model was developed using a CAD tool in the *Solid Works* program and the measurements used to build the parts were approximate, but adapted to suit the project. With the result obtained, some relevant factors were identified which should be studied in the future, such as the force applied to the piston head and the stress suffered by the connecting rod at the moment of explosion. As the study is only aimed at building a model that achieves greater torque with the same fuel consumption, these details, as well as the exact measurements of the parts, were not taken into account.

Keywords: Maximum torque. Mobile compression camera. Compression symbol.

1 INTRODUCTION

Engines, or power units, have been improved since they were first introduced around 1850, when the first patents began to appear. Their appearance put an end to the use of steam engines, due to their versatility, efficiency, low weight per horsepower and fast initial operation. After the appearance of these models, the first internal combustion engine was created by Belgian Jean Joseph Étienne Lenoir in 1860. He used the concept of Italian engineers Eugenio Barsanti and Felice Matteucci to create a two-stroke engine using explosive gas. This concept consisted of using the energy of expanding gases released by the combustion of an explosive mixture of air and hydrogen to transform the linear movement of a piston into a rotary movement, using a crankshaft. (BELLI, 2013).

According to Belli (2013), in 1861 Nikolaus August Otto and Langen built the first engine that compressed a mixture of air and gas for lighting. This engine had an electric spark ignition and was based on Lenoir's creation. The following year, the Frenchman Beau de Rochas based himself on Otto's study and published theoretical studies that established some thermodynamic principles. Otto, in turn, used Rochas' study and developed the Otto cycle engine in 1872, which used gasoline or coal gas as fuel. Years later, Otto cycle engines took to the streets and were used in vehicles.

Years later, in 1893, Rudolf Diesel proposed a new engine with ignition of the air-fuel mixture by very high compression. It came to be known as the Diesel engine and, the following year, he obtained his first patent for the proposed engine. In 1894, the first in-line 4-cylinder engine equipped with a camshaft and carburetor was presented by the Daimler Motoren Gesellschaft (BELLI, 2013).

Over the years, these developments have made engines bigger, more powerful and heavier. Later, studies were carried out into the materials used to manufacture them, which made them lighter and smaller, while maintaining or even increasing their power. Today's engines follow the lines of the Otto and Diesel engines, with some different characteristics. The Otto cycle uses low volatile fuels such as gasoline and alcohol and requires an electric spark for ignition. The Diesel cycle uses diesel oil as fuel. Compression and the consequent rise in air temperature cause combustion to occur when the fuel is injected into the chamber in pulverized form.

Due to the growing innovations in internal combustion engines, developing a model that improves diesel cycle engines in order to obtain more torque was the motivation behind this work. With a proposed change in combustion time, it is hoped that this objective can be achieved with the same fuel consumption. By installing a movable piston, compression will be maintained until the piston reaches a Midpoint (MP), between the Upper Deadpoint (UDP) and the Lower Deadpoint (LDP).

When the PM is reached, compression is increased so that the air-fuel mixture admitted at the start of the cycle is combusted.

Without taking into account the thermodynamics and exact measurements of a diesel engine, the parts were reproduced in a way that suited the project. Based on real values, we tried to idealize the size of the components so that they were as close as possible. All the parts were designed using a CAD tool and then assembled in such a way as to obtain the desired model for the project.

2 LITERATURE REVIEW

This topic will cover some of the fundamental characteristics of an internal combustion engine when it comes to torque. With a brief description of internal combustion engines, followed by a description of a diesel cycle engine, we'll cover the following points: engine times, what cylinder capacity is, some types of combustion chambers, compression ratio, power, torque and the performance that can be achieved with the change proposed in the project.

2.1 Internal combustion engines

The purpose of internal combustion engines is to transform chemical and physical energy into thermal energy, which in turn is transformed into mechanical energy. This mechanical energy is transmitted to the axle of a vehicle, allowing it to move. These types of engines can be classified into two main groups: the Otto cycle, created by the German Nikolaus August Otto, and the Diesel cycle, created by the engineer Rudolf Christian Karl Diesel (BOSH *apud* FERNANDES, 2012, p. 5). Both engines have very distinct characteristics, but we will only focus on the Diesel cycle in this work.

2.2 Characteristics of diesel cycle engines

Specific details about diesel cycle engines, the type chosen for optimization in this work, will be described below.

2.2.1 Diesel engine description

Diesel cycle engines are used in cases where greater power is required, without the need for high speeds, as they achieve maximum torque at low speeds. They are therefore used in large vehicles, especially those transporting heavy loads. They are not very applicable to passenger cars, due to their high pollution levels and sulphur concentration, and their use is aimed at the road market. Its advantage over an Otto cycle engine is that it doesn't need a spark plug for combustion to take place. This is due to the physical/chemical characteristics of diesel, which ignites only through compression (BOULANGER *apud* FERNANDES, 2012, p. 5).

Figure 1 shows the main components of a motor, which will be described later.

Figure 1: Schematic illustration of a motor.

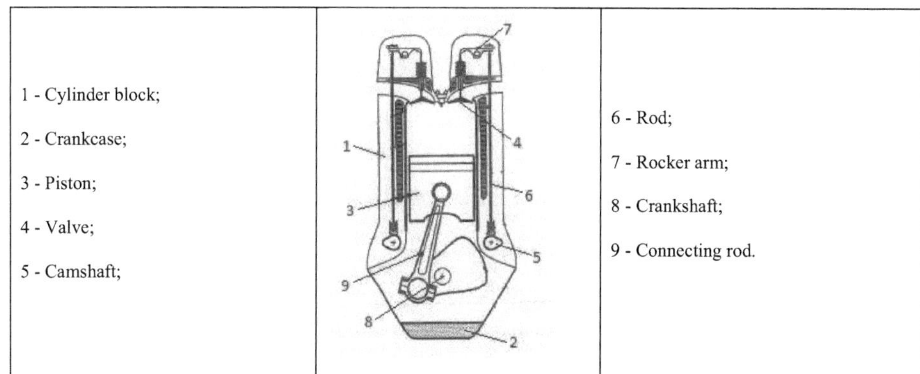

1 - Cylinder block; 2 - Crankcase; 3 - Piston; 4 - Valve; 5 - Camshaft;		6 - Rod; 7 - Rocker arm; 8 - Crankshaft; 9 - Connecting rod.

Source: MARTINS, 2013, p. 100. Adapted.

The following is a description of the parts of the engine, according to Martins (2013, p. 101 to 121):

1 - Cylinder block: a structure, usually made of cast iron, with cylindrical cavities in which the pistons work;

2 - Sump: lower cover responsible for storing oil;

3 - Piston: responsible for compression, receiving gas pressure and transmitting it to the crank-rod system. The piston rod has 3 rings, housed in the grooves, which are determined as follows: fire arrester, which has the function of cutting off the combustion flame, compression and oil scraper, which has the function of preventing the lubricating oil present in the crankcase from passing into the upper part of the cylinder;

4 - Valve: has the function of allowing the air or mixture to enter the cylinders and the burnt gases to exit. In general, intake valves have a larger head than exhaust valves because the pressure at the inlet is lower than at the outlet;

5 - Camshaft: shaft responsible for controlling the opening and closing of the valves. In 4-stroke engines, for every 2 turns of the crankshaft, the camshaft completes 1 turn;

6 - Rods: responsible for moving the rocker arm;

7 - Rocker arm: responsible for moving the valves;

8 - Crankshaft: responsible for receiving the reciprocating motion of the piston via the connecting rod and transforming it into the rotary motion of the engine shaft. It has elements such as: journals, supports, lubrication holes and counterweights;

9 - Connecting rod: its function is to transform the reciprocating movement of the piston rod into rotational movement of the crankshaft. It consists of a foot (the part that connects to the piston rod), a body and a head (the part that connects to the crankshaft).

2.2.2 Diesel 2-stroke engines

In 2-stroke engines, intake and exhaust occur at the same time, while compression and combustion occur at a different time.

In the first stage, the piston moves downwards. While air is being admitted through the intake window, it pushes the gases generated by the previous burn through the exhaust valve.

In the second half, the piston moves upwards, while the intake and compression windows close and the air is compressed. Just before the piston reaches top dead center, a mixture of fuel and lubricating oil is injected into the combustion chamber. This mixture ignites due to the high temperature and pressure, causing combustion.

The pressure caused by this combustion forces the piston downwards, generating work and ending the cycle after a complete turn of the crankshaft. The entry of the new mixture generates a "wash" inside the engine while expelling the gases from the previous burn.

2.2.3 4-stroke diesel engines

The Diesel cycle internal combustion engine has a different timing for each movement of the piston: intake, compression, combustion and exhaust (FIG. 2).

Figure 2: Timing in a diesel engine.

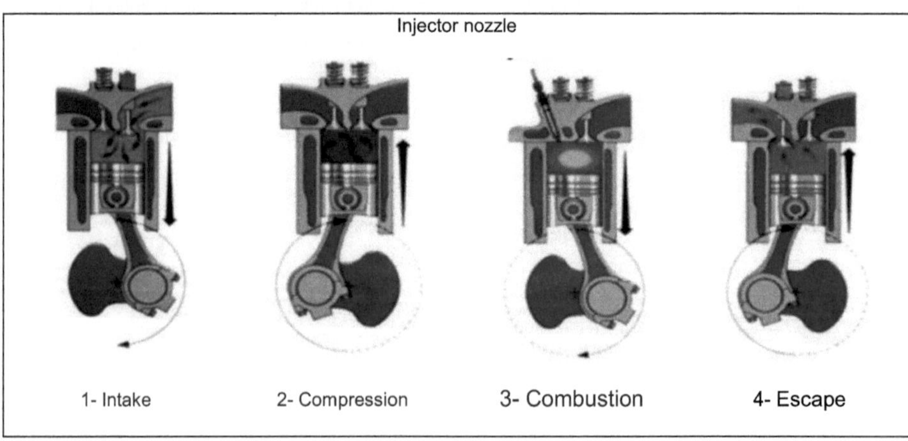

Source: Coutinho, 2016.

In intake, the piston is moving downwards, admitting air through the intake valve while the exhaust valve is closed.

During compression, both valves are closed and the piston moves upwards, compressing the air. Just before the piston reaches top dead center, fuel is injected into the combustion chamber. The heated and compressed air comes into contact with the finely pulverized fuel and ignites, initiating combustion.

During combustion, the piston makes a downward movement, driven by the expansion force of the burnt gases, while the intake and exhaust valves are closed. The force of the expansion of the gases is transmitted to the connecting rod via the piston rod, which in turn transmits this force to the crankshaft. This expansion causes the engine to rotate, making this time the only time that produces energy, while the others consume part of this energy.

Once combustion is complete, the piston moves upwards and expels the burnt gases inside the cylinder. At this point, the exhaust valve is open, while the intake valve is closed.

In a 4-stroke engine, the cycle ends every 2 complete turns of the crankshaft. Engines with 4 or more cylinders have an important characteristic in terms of cylinder coordination. In a 4-cylinder engine, the end pistons move in one direction, while the two central pistons move in the opposite direction. In this way, with each turn of the crankshaft, there are two combustions taking place and, with this synchronism in the movements, it facilitates the balancing of the engine (RACHE, 2004, p. 56).

2.2.4 - Engine capacity

Cylinder capacity is the volume displaced by the piston from top dead center to bottom dead center, multiplied by the number of cylinders in the engine (EQ. 1).

$$V = \frac{\pi * D^2 * h * n}{4} \tag{1}$$

where:

V = cylinder capacity [cm^3], [l] or [in];^{3}

D = piston diameter [cm] or [in];

h = track travel from top dead center to bottom dead center [cm] or [in];

n = number of cylinders.

2.2.5 Types of combustion chamber

The combustion chamber, or compression chamber, is where the air-fuel mixture burns in an engine. It is designed in such a way as to facilitate the combustion process, with the aim that the mixture is quickly and completely burnt each cycle.

There are different types of chamber design in diesel engines, which can be divided into two main groups: direct injection and indirect injection. Injection is called direct when it takes place directly on the piston. In indirect injection, the air enters a combustion pre-chamber or turbulence chamber, making rotary movements. This pre-chamber partially burns the fuel, which is then injected at the end of the compression period. The pressure of the combustion gases is gradually increased so that the mixture is not completely combusted beforehand (MARTINS, 2013, p. 328 to 331).

2.2.6 Compression ratio

According to Rache (2004, p. 63) and Martins (2013, p. 326), the compression ratio is the relationship between the initial volume of the cylinder, when the piston is in the lowest position (PMI), and the final volume of the cylinder, when the piston is in the highest position (PMS). When the piston is at PMS, we have the volume of the combustion chamber. This ratio varies between 16:1 and 18:1 and can be as high as 22:1 in diesel engines.

Let's take a six-cylinder Cummins 6.7 engine with 6,690 cylinders as an example. Each cylinder has a volume of 1,115 cm^3 (V), which refers to the total volume of 6,690 cm3 divided by the number of cylinders, when the piston is in the lowest position. The volume of the combustion chamber, when the piston is in the highest position, is 73.355 cm3 (v) (DODGE, 2017). The compression ratio can be calculated from Equation 2, which in the case of the example will be 16.2:1, i.e. the mixture is compressed into a space 16 times smaller than the initial one.

$$R_c = \frac{V + v}{v} \tag{2}$$

where:

Rc = compression ratio [dimensionless];

V = individual cylinder capacity (of a piston) [cm^3], [l] or [in];[3]

v = combustion chamber volume [cm3], [l] or [p p].0

2.2.7 Power

According to Varella (2010, p. 1 to 4), power can be: theoretical (estimated on the basis of physical properties and fuel consumption, where all the thermal energy from combustion is converted into mechanical energy), indicated (estimated on the basis of expansion pressure, dimensional characteristics and engine crankshaft rotation) or effective (estimated as a function of torque and engine flywheel rotation). For the purposes of this study, we will look at effective power, where this estimate is based on the principle of mechanical energy resulting from a tangential force on a circle of radius r.

Point P1 rotates around point P0 with angular velocity ω. Let's see, with the help of Figure 3, how the power calculation would be done.

Figure 3: Diagram of the tangential force applied at a distance r from the crank shaft.

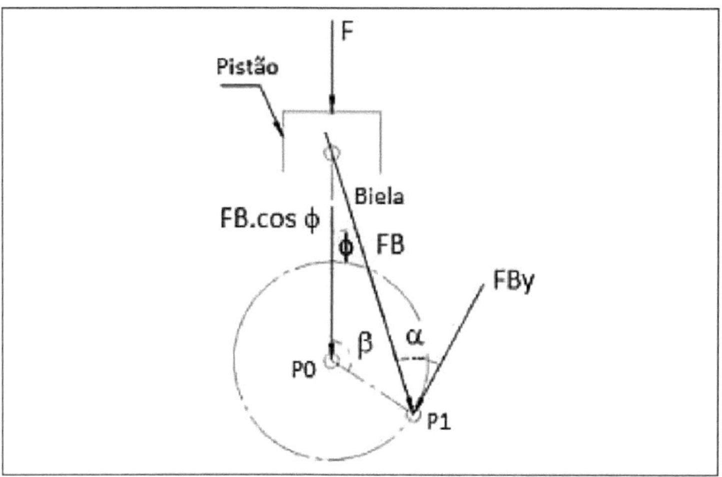

Source: VARELLA, 2010. Adapted.

The force acting on the rod can be obtained from Equation 5:

$$F = F_B * \cos(\emptyset) \tag{3}$$

$$F_B = \frac{F}{cos(\emptyset)} \tag{4}$$

$$F_B = \frac{P * A}{cos(\emptyset)} \tag{5}$$

where:

FB = force on the connecting rod, [N];

P = pressure at expansion, [Pa]:

A = area of the piston head, [m];^{2}

0 = angle between connecting rod stem and vertical [degrees].

2.2.8 Torque

In an engine, torque is the pressure exerted on the piston head area,

multiplied by the perpendicular distance from the axis to the direction of this force. With the help of Figure 4, you can understand this better.

Figure 4: Force applied at a point P at a distance r from the center of the crankshaft.

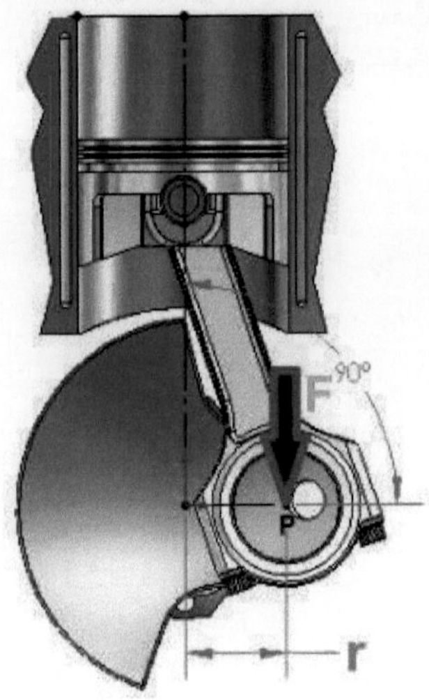

Source: DANTAS, 2010. Adapted.

Torque is defined as a force multiplied by a lever arm,

it follows that (EQ. 6):

$$T = F * r \tag{6}$$

where:

T = torque, [N.m];

F = acting force, [N];

r = distance between P and the origin, [m].

Figure 5 helps us to understand the relationship between the acting force F, the distance r and the inclination of the lever arm formed by the distance r.

Figure 5: Example of force application

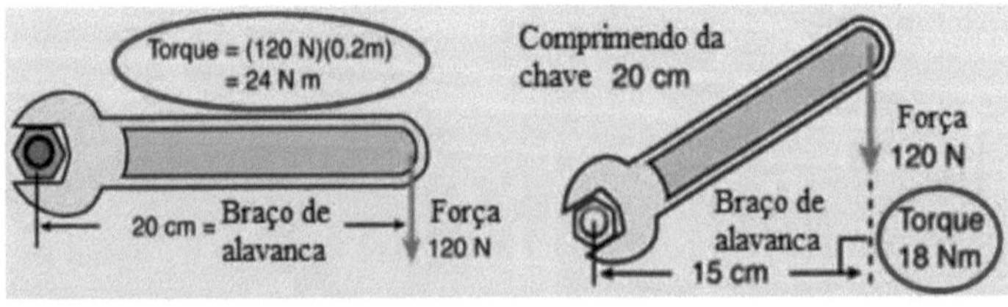

Source: NAVE, 2016. Adapted.

This means that the closer β is to 90°, the greater the torque provided. Thus, if β = 90°, we have the maximum possible torque. According to Varella (2010, p. 5), we can estimate the effective power in relation to the torque T and the angular velocity ω of point P, as shown below (EQ. 10):

$$P_E = F * \omega \qquad (7)$$

$$P_E = F * 2 * \pi * r * n \qquad (8)$$

$$P_E = \frac{T}{r} * 2 * \pi * r * n \qquad (9)$$

$$P_E = 2 * \pi * T * n \qquad (10)$$

where:

P_E = effective power, [W];

n = crankshaft speed (engine speed), [rps].

2.2.9 Income

Diesel cycle engines have a high efficiency when compared to Otto cycle engines, due to the type of fuel used and the working speed range in each. In Otto cycle engines, the fuel is injected together with the air, causing a loss during the mixing process. In the diesel cycle, the fuel is sprayed onto the compressed air at the end of compression. This makes better use of it, resulting in better performance.

The efficiencies used as parameters for comparing engines are: thermal, mechanical and thermomechanical. They are described by the following equations (VARELLA, 2010, p. 6 to 8):

a) Thermal efficiency is the percentage of thermal energy that is being converted into mechanical energy in the pistons. It can be described according to Equation 11.

$$R_T \quad \frac{P_I}{P_T}$$

where:

RT = thermal efficiency;

PI = indicated power, [W] PT = theoretical power, [W].

b) Mechanical efficiency is the percentage of mechanical energy in the pistons that is being converted into mechanical energy in the engine flywheel. It can be described according to Equation 12.

$$R_M \quad \frac{P_E}{P_I} \qquad 12)$$

where:

RM = mechanical efficiency;

PI = power indicated, [W];

PE = effective power, [W].

Thermomechanical efficiency is the percentage of thermal energy that is being converted into mechanical energy in the engine flywheel (EQ. 13).

$$R_{TM} \quad \frac{P_E}{P_T} \tag{13}$$

where:

MTR = thermomechanical efficiency;

PT = theoretical power, [W];

PE = effective power, [W].

According to Wong (*apud* Fernandes, 2012, p.7), the consumption of an engine at low speeds is higher than at intermediate speeds. This is because economic speeds correspond to the speeds with the lowest specific consumption values. Diesel cycle engines are therefore mainly used in heavy vehicles and cargo transportation because they have greater torque and high power at low and medium speeds.

3 METHODOLOGY

To develop the proposed project, the Diesel cycle model was used, due to its characteristics, and the *SolidWorks* program was used to reproduce the components of this model.

The main idea of the proposed project is based on the insertion of a movable compression piston that will maintain the compression of the mixture up to a certain point in the cylinder. When it reaches the Upper Dead Point (UDP), the point at which the piston compresses the mixture, the compression begins to lengthen. At this point the piston will begin to descend, gradually increasing compression until the point where the crankshaft forms a 90° angle with the tangential force exerted by the connecting rod (FIG. 6). At this point there will be sufficient compression for combustion to take place and a greater torque to be extracted from the engine. A number of models were proposed for this purpose, which were soon discarded.

Figure 6: The force applied at different times: a - Combustion in a standard engine; b - Combustion in an engine with a movable compression chamber.

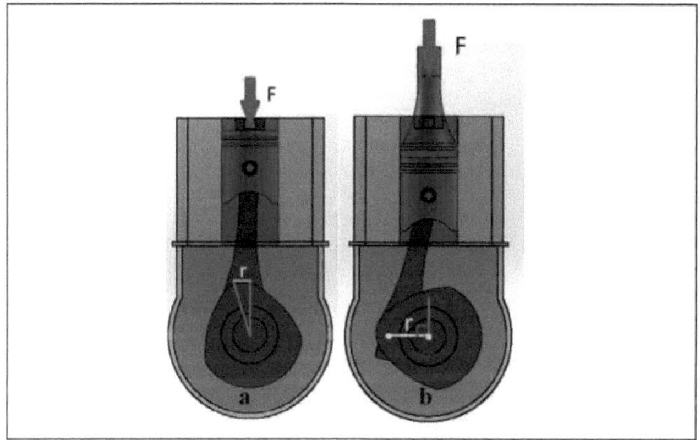

Source: Author.

Initially, the idea of reproducing a 4-cylinder engine was considered. Due to the difficulty of aligning the timing of the camshaft, crankshaft and movable compression piston, it was decided to reproduce just one cylinder.

One of the models consisted of an electro-pneumatic control that would move the piston with the help of an alternative system. This system would emit electrical pulses to drive the piston, which would then inject air into an auxiliary chamber. However, this model was discarded due to the fact that air is a compressible fluid, which would result in a loss of compression inside the cylinder.

Another model proposed was an electro-hydraulic one which, with the help of a non-compressible fluid, would drive the piston. On the other hand, the electric drive for injecting and withdrawing the fluid would not be fast enough to keep up with the high speeds of an Otto cycle engine.

A third model proposed was the same electro-hydraulic system for a diesel engine, which works at lower speeds than an Otto cycle. However, even at lower speeds, the injection and withdrawal of fluid into the engine cylinder would be a problem, due to the speed with which this would occur.

After systematic analysis of these models, it was concluded that the best for this study would be a mechanism with a mechanical drive. In this model, the piston would move by means of a system similar to the intake and exhaust valves. To do this, it would be necessary to adjust the timing and find the ideal measurements for each component. After a series of trials and errors, an apparently ideal model was arrived at.

The motor was assembled by estimating the dimensions of all its components and then positioning each one. The engine was then put into operation and any errors were checked. Then, the dimensions of the components were corrected, the motor was put back into operation and the steps were repeated until a satisfactory model was achieved.

Section 3.1 explains how each component was designed and reproduced in *SolidWorks* for the final model.

3.1 Modeling the pieces

The modeling of the components used to develop the prototype was based on approximate sizes of vehicle engine parts, based on books. However, a reduction scale was not used for this, as this would make it easier to build and adjust them.

Table 1 lists the parts built to assemble the model, with the main measurements for each one.

Table 1 - Parts designed to develop the model.

Track: the track was made with a circumference of 50 mm in diameter and a height of 62.50 mm. The ring grooves were made in the head and the 12.50 mm diameter holes were drilled in the center of the body of the piston pin (Appendix A).	

Block: the block was built 120 mm high, 130 mm wide and 120 mm long. Three holes were drilled in the top face, two of which were 25 mm and the third 50 mm (Appendix B).	
Crankcase: The crankcase had a total height of 149.50 mm, a width of 120 mm and a length of 130 mm. The central hole was the same as the crankshaft, 30 mm in diameter (Appendix C).	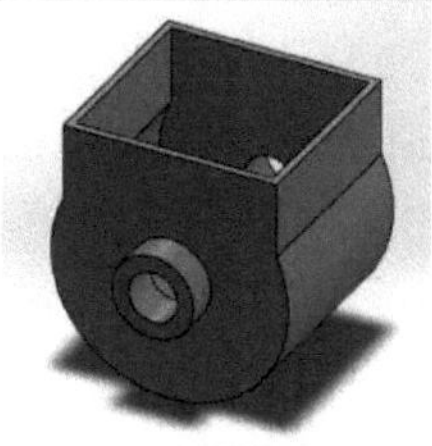
Crankshaft gear 1: Gear 1, which will be coupled to the crankshaft, was made with a diameter of 60 mm and 16 teeth, each 6 mm high. The central hole was the same as on the crankshaft, 30 mm in diameter (Appendix D).	
Crankshaft gear 2: Gear 2, which will also be coupled to the crankshaft, was built with a diameter of 83 mm and 23 teeth, each 6 mm high. The diameter of the central hole was the same as that of the crankshaft, with a diameter of 30 mm (Appendix E).	

Piston camshaft gear: this gear receives the movement of gear 2 and transmits it to the piston camshaft. It has been made with a diameter of 166 mm and 46 teeth, 6 mm high. The central hole, which will engage it, is 60 mm in diameter (Appendix F).	
Camshaft gear: this gear will receive the movement from gear 1 and transmit it to the camshaft. It is 120 mm in diameter and has 32 teeth, each 6 mm high. The central hole, which will engage it, is 20 mm in diameter (Appendix G).	
Connecting rod: the foot of the connecting rod is 25 mm outside diameter and 12.50 mm inside diameter. The head was made with 42 mm outside diameter and 32 mm inside diameter. The distance between the center of the foot and the center of the head was 120 mm to form the connecting rod body (Appendix H).	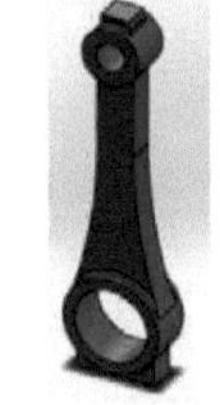
Piston rocker arm: the rocker arm that will drive the piston was built with a total length of 132 mm, running from the center of one end to the other. The diameters of these ends are 15 and 11 mm, and they are 47 and 85 mm from the center, respectively. The central hole in the rocker arm has an external diameter of 18 mm and an internal diameter of 15 mm (Appendix I).	

Valve rocker arm: the rocker arm that transmits the vertical movement to the valves was made 97 mm long. The ends were 10 and 11 mm in diameter, 37 and 60 mm away from the central hole in the rocker arm (Appendix J).	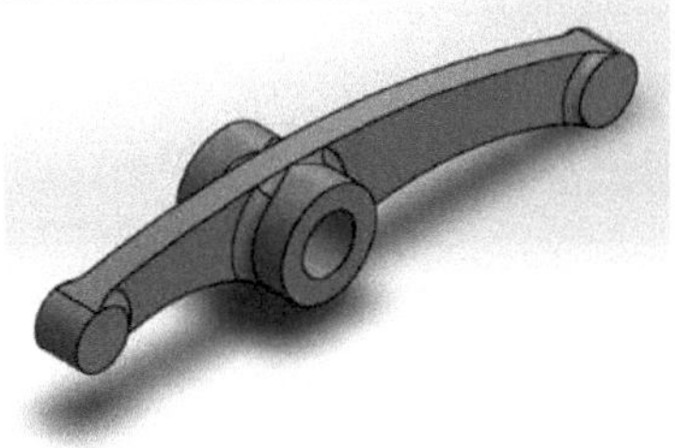
Variable compression piston rod: the rod that transmits the movement of the cam to the rocker arm was made 225.33 mm high. The diameters of the base, body and head were 25, 12 and 20 mm respectively (Appendix K).	
Valve rod: The rod that will carry the movement from the cams to the valve rocker arms was made with a total height of 257.50 mm. The base, body and head were 20, 8 and 12 mm in diameter respectively (Appendix L).	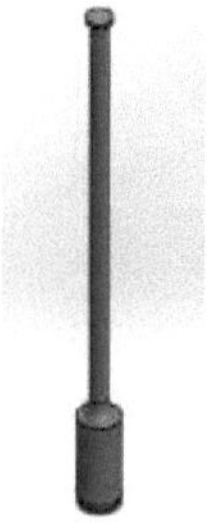
Valves: The valves were made with a height of 90 mm, a base diameter of 25 mm and a body diameter of 5 mm (Appendix M).	

Crankshaft: the crankshaft was made 230 mm long, with the shafts 30 mm in diameter and the counterweights 12 mm wide. The trunnion was 32 mm in diameter and 30 mm long (Appendix N).	
Piston rocker arm bearing: the bearing that will hold the piston rocker arm was made with a total length of 228.50 mm and a diameter equal to the central hole of the rocker arm, 10 mm (Appendix O).	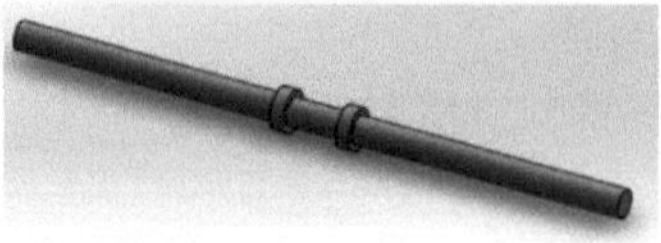
Valve rocker arm bearing: the bearing that will support the valve rocker arms was made 228.50 mm long and 10 mm in diameter. A central protrusion with an increase of 4 mm in diameter and 64 mm in length was made to separate the two rocker arms (Appendix P).	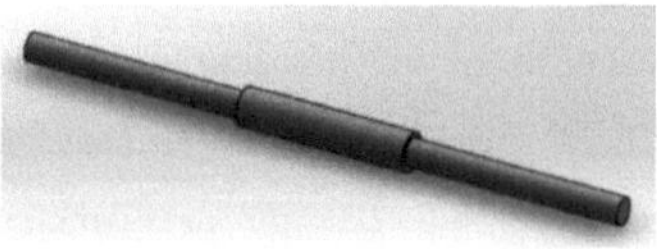
Pin: the pin was made as a hollow tube 50 mm long. The outer diameter is 12.50 mm and the inner diameter is 7.50 mm (Appendix Q).	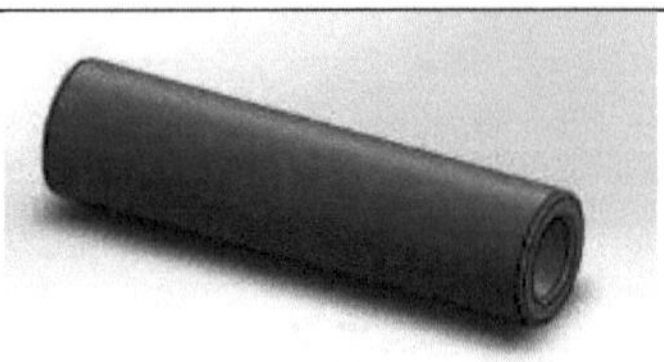
Mobile compression piston: the piston that will compress the air was made with a height of 88 mm, a base diameter of 50 mm and a top diameter of 20 mm (Appendix R).	

Piston cam shaft: the cam shaft that moves the piston was made with a diameter of 60 mm and a length of 228.50 mm. The cam is 20 mm wide, has a base diameter of 85 mm, a "nose" diameter of 17 mm and radii (flanks), larger and smaller, of 200 mm and 85 mm respectively. The height from the base to the nose was 101 mm.	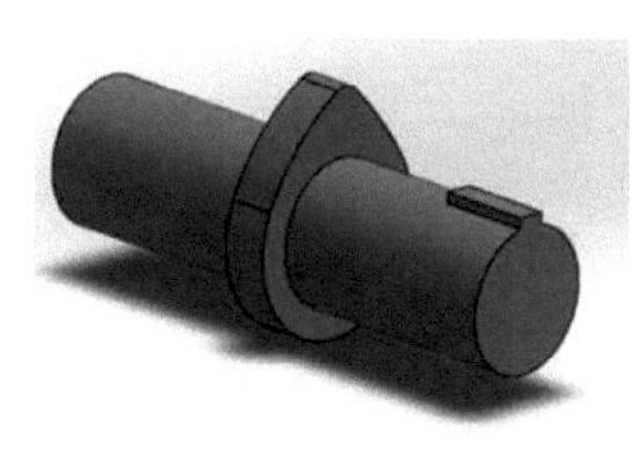
Valve camshaft: the camshaft was made with a diameter of 20 mm and a length of 228.50 mm. The cam was built with a 24 mm diameter base, a 5 mm diameter "nose" and radii (flanks), larger and smaller, of 40 mm and 16 mm respectively. The height from the base to the "nose" was 29 mm.	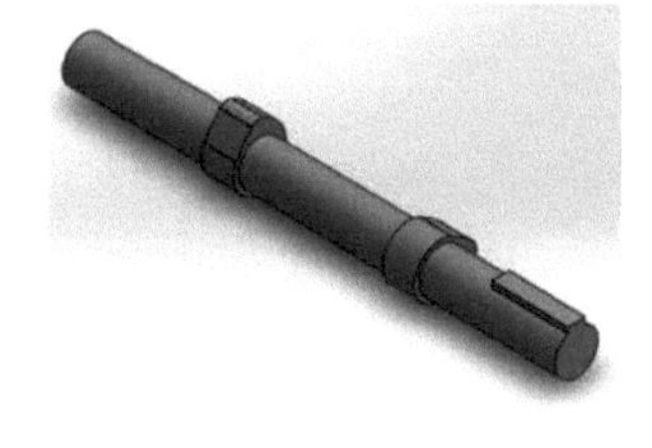

Source: Author

The size of the pieces was not the main focus of the project, but it was based on books and catalogs to reproduce them. In addition, their measurements were altered to fit the prototype. All the components were easily reproduced, with the exception of the cam, which is responsible for the movement of the piston. This cam was modeled through trial and error until an ideal size was found that fit well with the other components.

3.2 Assembly

After reproducing each part in the computer program, the engine was assembled, as described below.

Once all the engine components had been created, assembly began. The first part inserted was the crankcase, which will serve as a reference for all the other components. Its origin is fixed in the center, as shown in Figure 7.

Figure 7: Carter origin.

Source: Author.

After attaching the crankcase, the engine block was inserted and, using the advanced width positioning feature, they were centered. They were then fixed with the standard coincident positioning (FIG. 8).

Figura 8: Positioning the block in the sump.

Source: Author.

The next component inserted was the crankshaft. The transparency of the crankcase was altered to make it more visible, since the crankshaft will be inside it. Using the standard concentric positioning feature, the shaft was positioned next to the crankcase holes and, using the advanced positioning by width, the crankshaft was centered (FIG. 9).

Figura 9: Positioning the crankshaft in the crankcase.

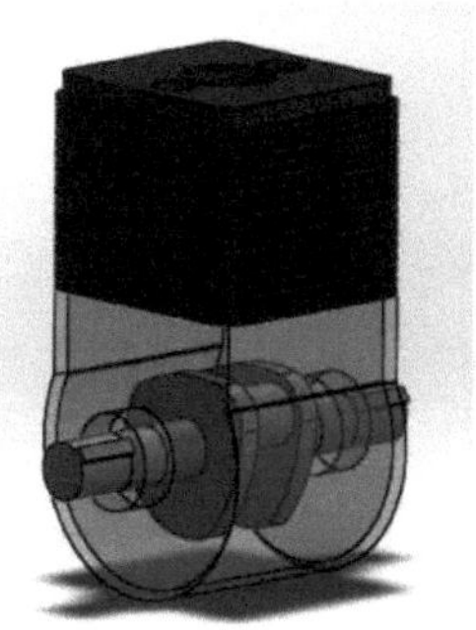

Source: Author.

After positioning the crankshaft, the gears that will transmit movement to the camshafts were inserted. With standard positioning, they were positioned concentrically to the crankshaft and, with coincident standard positioning, they were fixed to the keys (FIG. 10).

Figura 10: Positioning the transmission gears on the crankshaft.

Source: Author.

The next components inserted into the assembly were the camshafts for the piston and valves. They were positioned parallel to the crankshaft using construction lines. The center of the piston and camshafts were placed at a distance of 132 mm and 97 mm respectively from the center of the crankshaft. The key of the piston cam shaft was turned to the side of the crankshaft's largest gear. As a result, the key of the camshaft of the camshafts faced the smaller gear side of the shaft, as can be seen in Figure 11.

Figure 11: Positioning of the control axes.

Source: Author.

After positioning the shafts, the gears corresponding to each one were inserted. The largest gear, with 46 teeth, was positioned concentrically on the piston cam shaft and fixed with the positioning coinciding with the key. The same was done with the smaller 32-tooth gear, which was fixed to the cam shaft of the valves, as shown in Figure 12.

Figure 12: Positioning of the camshaft gears.

Source: Author.

For the next step, the block and crankcase were hidden to make it easier to insert the connecting rod, as can be seen in Figure 13. Its head, the largest hole, was positioned concentrically to the crankshaft wedge and, using advanced positioning by width, was centered on it.

Figura 13: Positioning the connecting rod.

Source: Author.

After positioning the connecting rod, the piston rod and pin were inserted into the assembly. The pin was positioned concentrically to the hole in the piston rod and, using the advanced width positioning feature, centered in relation to it. The pin was selected again and positioned concentrically to the foot of the connecting rod (smaller hole). To center the piston rod on the connecting rod, the standard coincident positioning feature was used, relating the frontal plane of the piston rod to a central point on the foot of the connecting rod. Finally, the piston body was positioned concentrically to the block's central hole (FIG. 14).

Figura 14: Positioning the piston and pin.

Source: Author.

With the right plane as a reference, two parallel planes were opened, where the rods that transmit movement to the piston and valve rocker arms were inserted. They were positioned 97 and 132 mm away from the reference plane, towards the camshaft of the valves and the camshaft of the piston, respectively. Vertical construction lines were made in these two planes, passing through the center of the cams, where they were positioned concentrically to the rods. Using mechanical cam positioning, the rods were placed on the corresponding cams, as shown in Figure 15.

Figure 15: Positioning of the control rods.

Source: Author.

To position the bearings that support the rocker arms, a vertical construction line 280 mm from the center of the crankshaft was made. Then, on top of this line, two other horizontal lines were made, 60 and 85 mm long, towards the camshafts of the valves and the piston, respectively. As shown in Figure 16, the bearings were positioned concentrically at the end of these horizontal lines and parallel to the end of the crankshaft.

Figure 16: Positioning of the bearings.

Source: Author.

Then, as can be seen in Figure 17, the valves and the variable combustion chamber symbol were inserted in a similar way. The elements were selected and positioned concentrically in the block: the symbol in the central hole and the valves in the side holes.

Figure 17: Positioning of the piston and valves.

Source: Author.

The last components inserted were the rocker arms that move the piston and valves. The central holes of the rocker arms were positioned concentrically, while the side walls were positioned coincidentally with the side walls of the bearings. The cylinders at the smaller ends were positioned tangent to the rods, while the cylinders at the larger ends were positioned tangent to the heads of the valves and the piston (FIG. 18).

Figure 18: Positioning of the rocker arms.

Source: Author.

Once all the components had been positioned, the movements were synchronized. The start of the explosion time was taken as the reference point and, when the piston is at PMS, this instant is 0°. The ratio between the largest gear on the crankshaft and the gear on the piston camshaft is 2 to 1. This means that for every two turns of the crankshaft, the piston camshaft completes one turn. Before finalizing the positioning, the crankshaft lever arm was moved close to the 90° angle and the cam of the piston camshaft close to the vertical position. This is the maximum point of descent of the piston, where the explosion will occur, as shown in Figure 19.

Figura 19: Positioning of gear 2 of the crankshaft with that of the camshaft.

Source: Author.

After positioning the piston camshaft with the larger crankshaft gear, the valve camshaft was positioned with the smaller crankshaft gear. The crankshaft lever arm was rotated to a position of 180°, while the camshaft was placed in a position where one of the valves started to open. In this way, when the piston starts to rise, the piston returns to its original position and then the exhaust valve opens. When the piston reaches PMS, the exhaust valve closes and the intake valve opens. Figure 20 shows the arrangement of the components after the gears have been synchronized.

Figura 20: Positioning of gear 1 of the crankshaft with the camshaft.

Source: Author.

Once assembled, the model was tested by turning it by hand to check whether the components would collide or act in an undesirable way. Once this stage had been completed, a video demonstrating the operation of the model was developed to clearly explain how the variable compression piston would work (FIG. 21).

Figure 21: Final engine design: a - with fin; b - without fin on engine block.

Source: Author.

4 RESULTS AND DISCUSSION

By analyzing the simulation of the model in operation, we can see the difference between the compression at two key moments (FIG. 22). In Figure 22.a, the lever arm is in a position of 0° and the piston is at PMS (Top Dead Center), with a volume of 19.62 cm^3 in the compression chamber. Near this point, in conventional direct injection engines, diesel would be injected for combustion to take place. Unlike what normally happens, in the proposed project, the air-fuel mixture would be admitted in the 0° position, but combustion would only take place in the 90° position. At this second moment, shown in Figure 22.b, the compression chamber has a volume of 16.30 cm3, 3.32 cm3 less and with the same amount of mixture. Working with hypothetical values and assuming that ideal compression was achieved at this moment, this would ensure that combustion occurred at the desired time. With this, it would be possible to achieve a higher torque than would normally be achieved, as seen in the literature reviews.

Figure 22: a - 0° position of the lever arm; b - 90° position of the lever arm

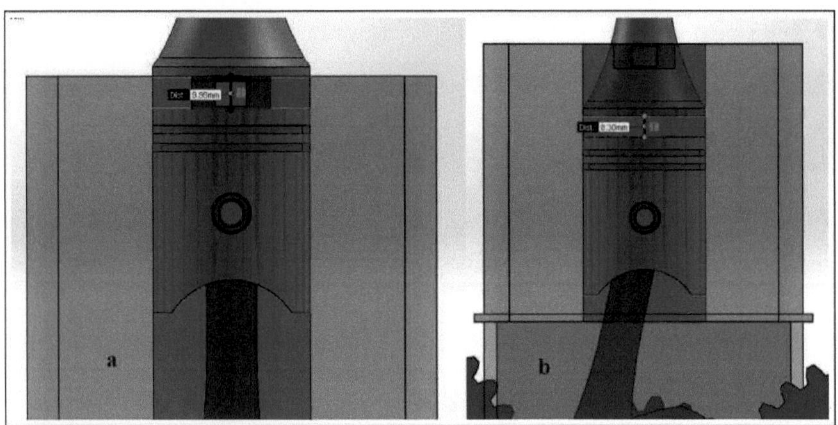

Source: Author.

The sequence of images (FIG. 23 to FIG.26) reproduce the times obtained after the changes made, simulating the complete engine cycle. The letters AD refer to the intake, while the letters ES refer to the exhaust.

Figure 23: a - End of exhaust time; b - Admission time in progress.

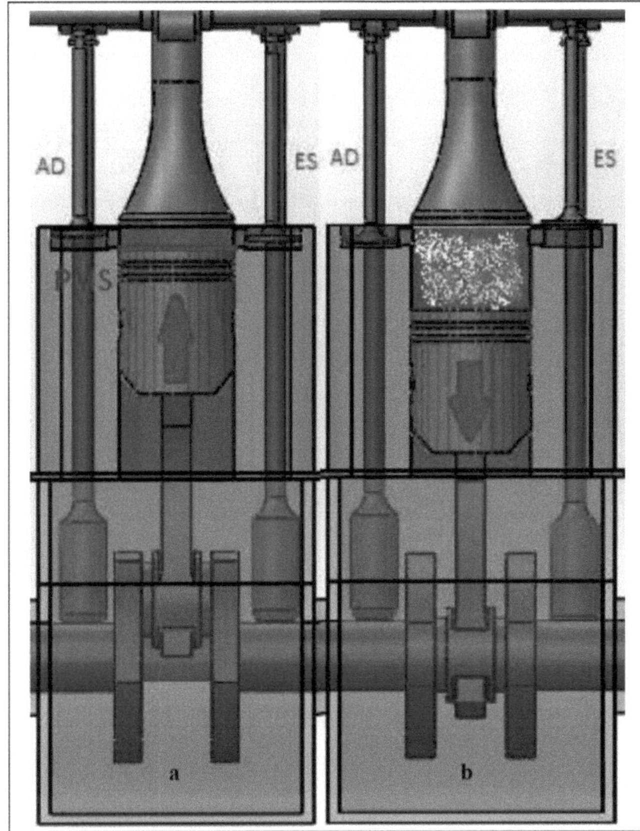

Source: Author.

The intake process starts from the PMS, when the intake valve opens and air-fuel enters the compression chamber, while the piston is moving towards the PMI (Lower Dead Center). In image 23.a, the piston is finishing the exhaust and starting the intake. In image 23.b, the piston is finishing the air intake time.

Figure 24: a - End of intake time; b - Compression time; c - End of compression time.

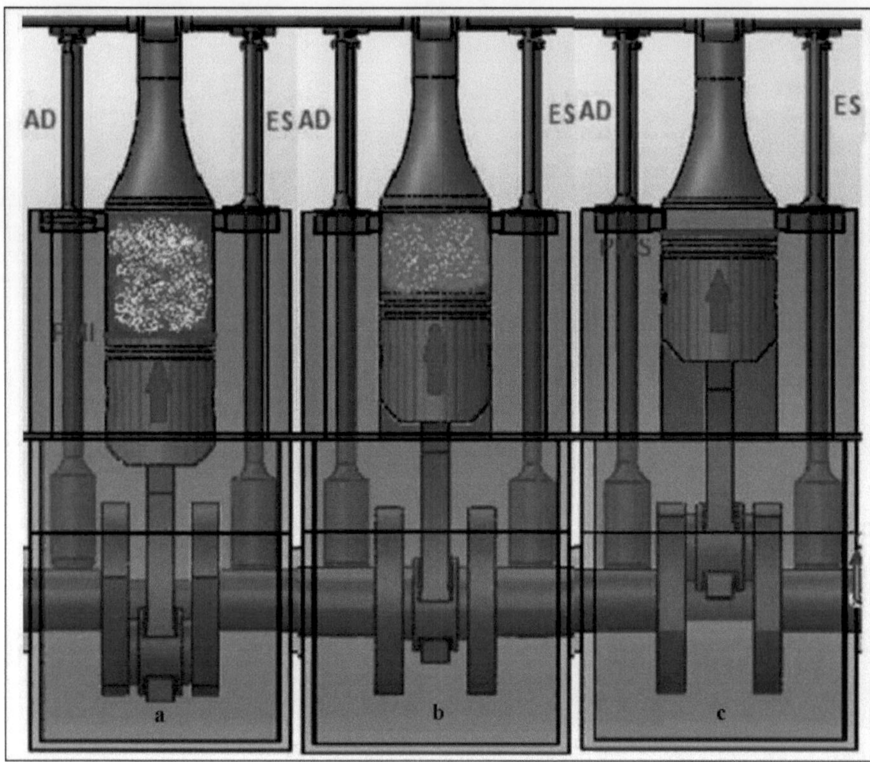

Source: Author.

In image 24.a, the piston is about to start its compression time. At this point, the piston will begin its upward movement and the intake valve will close. After the cylinder has admitted the mixture, with the intake valve already closed, the piston, starting at PMI, begins to compress this mixture towards PMS, as seen in image 24.b. After reaching the PMS (FIG. 24.c), the compression time ends, starting another time.

Figure 25: a - Lengthening of compression time; b - Explosion; c - Start of combustion time.

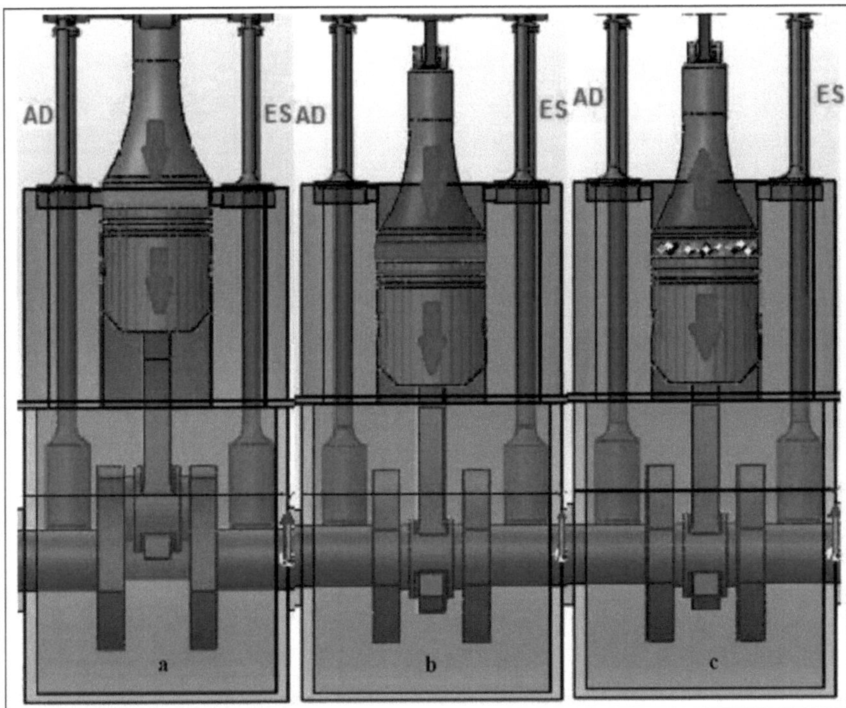

Source: Author.

Once compression is complete, a synchronized downward movement of the piston starts with the compression piston (FIG. 25.a). When both reach a midpoint, the chamber reaches a higher compression rate than that initially achieved, causing the explosion to occur (FIG.25.b). At this point, the crank arm force is forming a 90 degree angle with the crankshaft, and this is where the highest possible torque is obtained, as proposed in the design. Once combustion is complete (FIG. 25.c), the piston continues its downward movement, while the piston begins an upward movement, to end the explosion time.

Figure 26: a - Start of exhaust time; b - exhaust of gases.

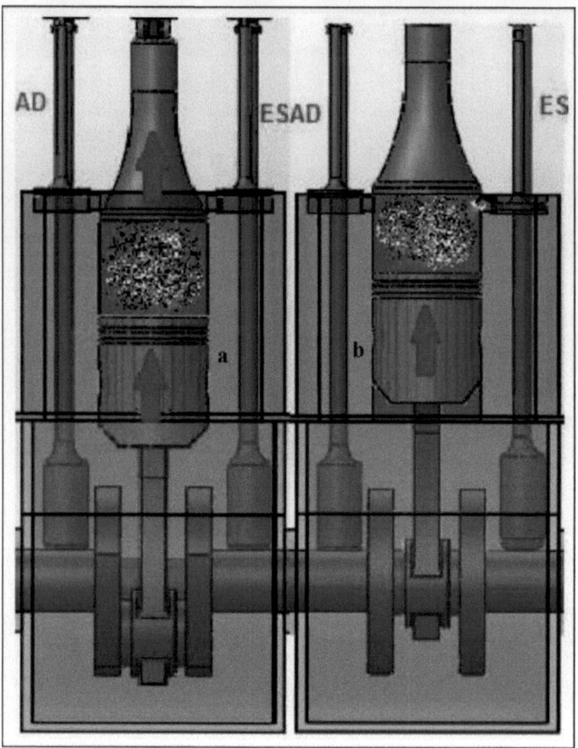

Source: Author.

To finish the cycle, the piston ends its upward movement, while the piston continues to rise (FIG. 26.a). When the piston returns to its initial position, the exhaust valve opens and the piston expels the gases from the burner, through its upward movement, ending one cycle and starting a new one (FIG. 26.b).

It can be seen in the sequence of images (FIG. 23 to FIG. 26) that the times are similar to those of a standard engine, differing only in the inclusion of the piston. With this inclusion, it can be considered that the compression has been lengthened in order to reach the ideal moment for combustion to take place.

With combustion occurring at the desired moment, when the crank arm forms a 90-degree angle with the crankshaft, the greatest possible torque can be extracted from an engine. This increase in torque is an advantage when you take into account the fact that the engine's volumetric capacity remains the same. As a result, you get more torque with the same fuel consumption.

Furthermore, as described in Section 2, the increase in torque will generate an increase in effective power which, in turn, will generate an improvement in thermal and thermomechanical performance.

But to confirm this, more in-depth studies are needed into the strength of materials and thermodynamics itself.

The Diesel cycle was chosen because of its low working speeds, compared to the Otto cycle engine, and the absence of a spark plug. This is because the mixture only needs compression for combustion to occur. In addition, the mechanical drive was chosen because it can be made by the same shaft as the camshaft or in a similar way on another shaft. In this way, it is hoped to achieve a higher torque than that obtained in a diesel engine, taking into account the angle formed between the lever arm and the crankshaft at the moment of combustion. With a higher torque, a higher effective power can be achieved and, consequently, a gain in mechanical efficiency and thermomechanical efficiency.

The changes presented in this study can be optimized in a standard engine due to the fact that the piston camshaft is independent of the other components. In addition, this control can be adapted to be activated only at times when the engine requires more torque, by means of a variable control[I].

I Variable control is a system responsible for varying the time the piston is actuated. It can be through opening variation, where a central electronic unit selects a different cam profile based on the load and speed of the engine, or it can be through phase variation, where an actuator turns the camshaft on its own axis, changing the timing of the piston.

5 CONCLUSION

Initially, models were proposed for the development of this study. Due to the unfeasibility of some of these models, it was concluded that a mechanical drive of the variable compression piston would be the most suitable for the project.

After numerous attempts to size the components so that they wouldn't collide, and the great difficulty of aligning the motor timing with the valve and piston actuation timing, the expected final result was achieved.

The biggest limitation of this project was finding the right size for the cam that drives the compression piston. Getting it to actuate at the right time, with the required amount of advance and at the ideal moment, were difficulties encountered when building the cam. A small change in the size of the cam meant a large variation in the piston drive command, hence the great difficulty.

Once the model had been designed, all the measurements were checked and the components were verified to ensure that they did not clash. After the studies, it was possible to obtain a final model that matched the initial proposal and was compatible with the main idea of the project. This is a single-cylinder diesel engine model, which would be capable of extracting more torque than a standard engine. In this work, only the central idea was presented, without a detailed analysis of the mechanical design of the parts (which would take strength into account). The CAD design only took into account geometric issues for visualization and a better understanding of the prototype.

For future studies, we recommend designing the components included in the model, comparing them with a standard motor and checking whether the torque gain would be effective. In addition, studies such as the increase in effective power, mechanical and thermomechanical performance could also be carried out to verify the feasibility of the project. A new model for the piston and the forces acting on the components that are directly linked to the force exerted by the pressure acquired at the moment of combustion are other proposals that could enrich and add value to this work.

6 REFERENCES

BELLI, M. **Internal Combustion Engines, A Brief History**. Available at: <http://www.autoentusiastasclassic.com.br/2013/03/motores-combustao-interna-uma- breve.html>. Accessed on: June 1, 2016

COUTINHO, M. Basic Operation of the 4-Stroke Diesel Engine. Available at: <https://www.linkedin.com/pulse/funcionamento-b%C3%A1sico-do-motor-diesel-4-tempos-marcos-coutinho>. Accessed on: September 19, 2017.

DANTAS, A. **R/L ratio: a graphical analysis**. Available at:

<http://www.autoentusiastasclassic.com.br/2010/09/relacao-rl-uma-analise-grafica.html>.

Accessed on: August 22, 2017.

FERNANDES, A. **Analysis of Diesel Engine Performance with the Use of Biofuel Obtained from the Reuse of Vegetable Oil**. Santa Bàrbara d'Oeste, 2012. Dissertation (Master's Degree in Production Engineering). Faculty of Engineering, Architecture and Urbanism, Methodist University of Piracicaba, UNIMEP;

Datasheet RAM 2500 LaramieCrewCab 6.7 Turbodiesel 4x4. Dodge. Available at: <www.dodge.com.br/content/dam/ramportal/pdfs/ficha_tecnica_ram_2500.pdf>. Accessed on: August 30, 2017.

MARTINS, J. **Internal Combustion Engines**. 4.ed. Sao Paulo: Saraiva, 2013.

Internal combustion engines. Available at:

<http://wp.ufpel.edu.br/mlaura/files/2013/01/Apostila-de-Motores-a-Combust%C3%A3o-Interna.pdf>. Accessed on: May 31, 2016.

RACHE, M.A.M. **Diesel Mechanics: Trucks - Pick-ups - Boats**. Sao Paulo: Hemus, 2007.

SILVA, E.da. **Electronic Injection of Diesel Engines: EDC, PLD, UI, Common Rail**. 1. ed. Sao Paulo: Jubela, 2006.

NAVE, C.R. **Torque**. Georgia State University, 2016. Available at: <http://hyperphysics.phy-astr.gsu.edu/hbase/torq.html>. Accessed on: August 22, 2017.

VARELLA, C. A. A. **Estimating the power of internal combustion engines**. UFRRJ, 2010.

7 APPENDICES

APPENDIX A - Track

Construction of the runway began with the front plane. First, sketch 1 was made, using the intelligent dimension command to determine the measurements, as shown in Figure A1. Still in sketch 1, the grooves where the rings are fixed and the arc that will form the head of the track were created.

Figure A1: a - sketch 1 of the track; b - drawing of the grooves.

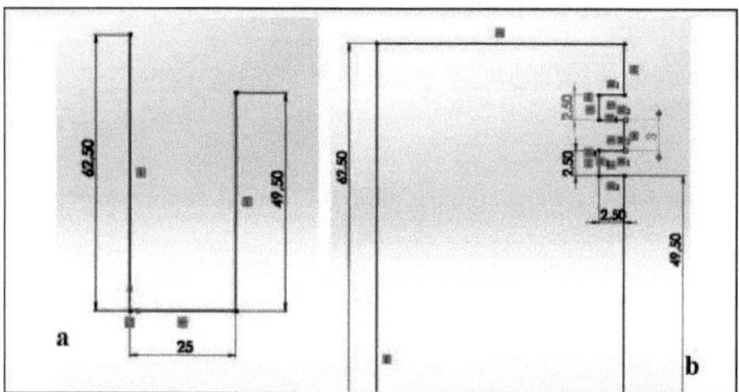

Source: Author.

Next, the whole of sketch 1 was selected and *offset* inwards by 2 mm. Then, using the extend entities and trim entities command, sketch 1 was as shown in Figure A2.

Figure A2: Sketch 1 of the finished runway.

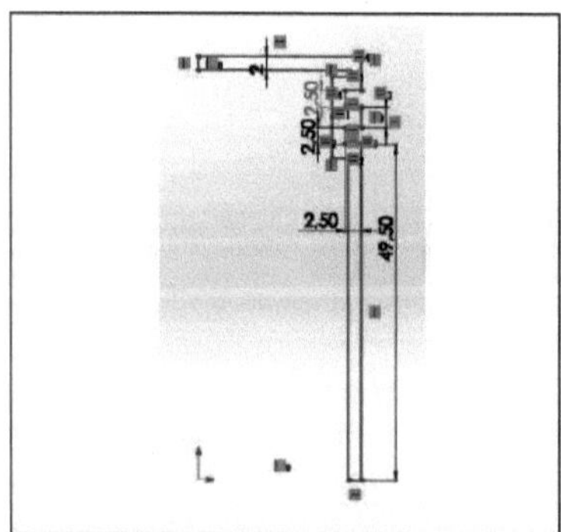

Source: Author.

The rebound/revolutionized base feature was used to give the track a cylindrical shape, as can be seen in Figure A3.

Figure A3: Revolution of the sketch 1 of the track.

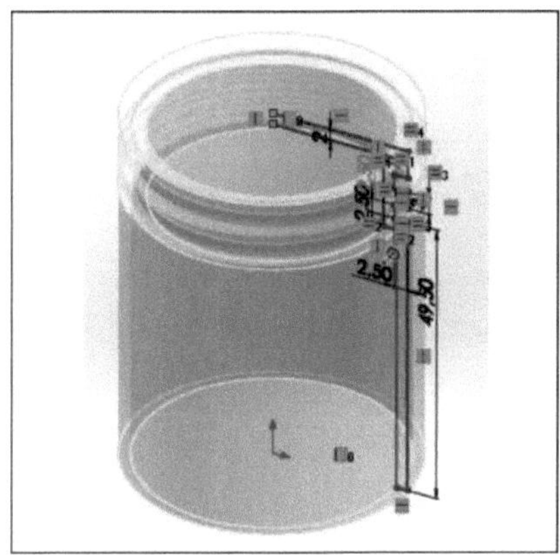

Source: Author.

Sketch 2 was made from the frontal plane to make the extruded cut of the lower part of the track, as shown in Figure A4.

Figure A4: Sketch 2 and extruded section of the track.

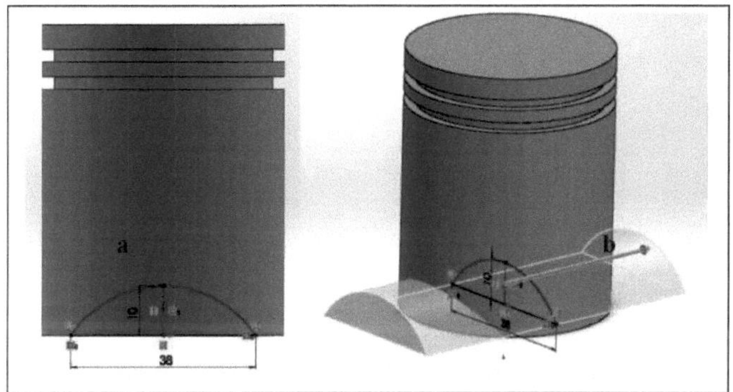

Source: Author.

The front plane was selected for the construction of sketch 3 and an extruded cut was made, through which the pin that will connect the piston pin to the connecting rod will pass, as shown in Figure A5.

Figure A5: Sketch 3 and extruded section of the track.

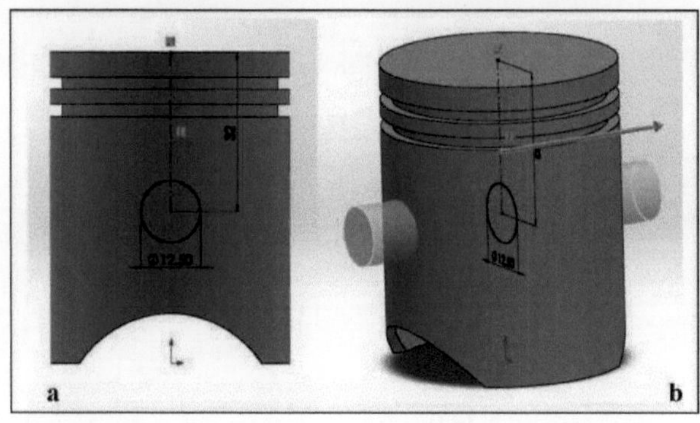

Source: Author.

Finally, the final shape and type of part material chosen, polished aluminum, can be seen in Figure A6.

Figure A6: Finished track.

Source: Author.

APPENDIX B - Engine block

Sketch 1 was first made from the top plane and then a 110 mm extrusion was applied to form half of the block. This stage can be seen in Figure B1.

Figure B1: a - sketch 1; b - block extrusion.

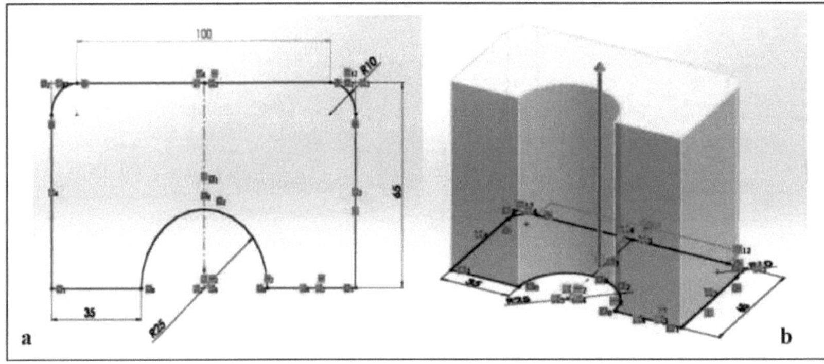

Source: Author.

Selecting the top face of sketch 1, sketch 2 was created using the *offset* command and extruded by 3 mm towards the other top face of the block. Using the linear pattern command, with a spacing of 6 mm and 17 instances, the block's cooling fins were formed, as shown in Figure B2.

Figure B2: a - sketch 2; b - extrusion of the block and formation of the fins.

Source: Author.

To form the other half of the block, the entire body was mirrored in relation to the front plane. On the upper face of the block, sketch 3 was created to form the holes through which the intake and exhaust valves will pass. Two 25 mm circumferences were made, in addition to a 10 mm extruded cut made into the block, as can be seen in Figure B3.

Figure B3: a - sketch 3; b - mirror image of the block and intake and exhaust ports.

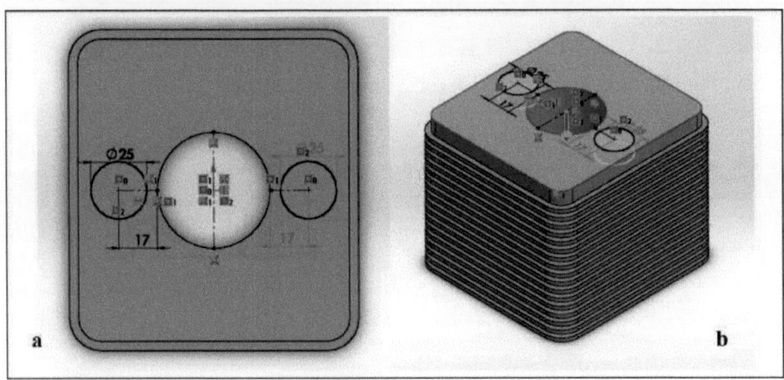

Source: Author.

From the front plane, sketch 4 was opened and a rectangle was made, 8 mm high and 10 mm wide. The center of this rectangle is located 6 mm from the top of the block, as can be seen in Figure B4. Next, a 90 mm long extruded cut was made in the middle plane to connect the valve bores to the cylinder.

Figure B4: a - sketch 4; b - connection of intake and exhaust bores to the cylinder.

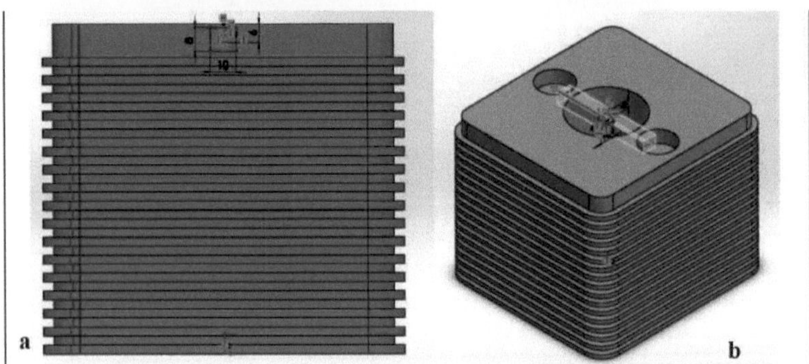

Figure B5 shows the type of material chosen, cast iron, and the finished block.

Figure B5: Finished block.

Source: Author.

To create the carter, sketch 1 was made from the frontal plane. Then a 20 mm blind extrusion was made in direction 1 and a 5 mm blind extrusion in direction 2, as shown in Figure C1.

Figure C1: a - sketch 1; b - and crankcase extrusion.

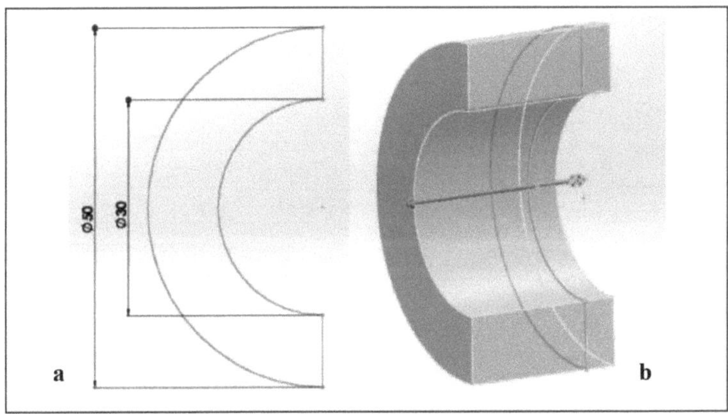

Source: Author.

Sketch 2 was created from the frontal plane to form the crankcase wall, which was extruded by 5 mm in the direction shown in Figure C2.

Figure C2: a - sketch 2; b - crankcase extrusion.

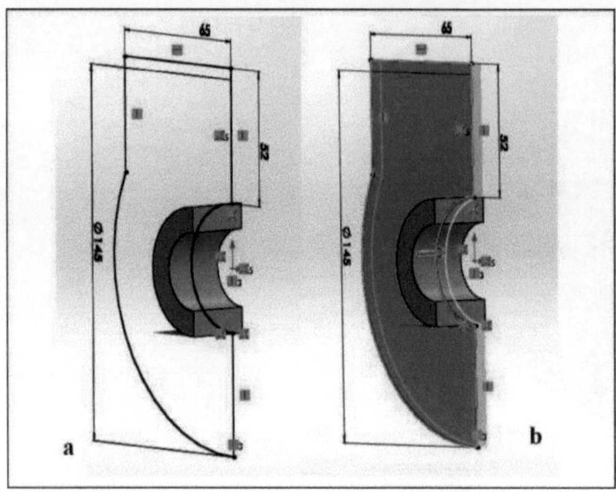

Source: Author.

A new plane 1 was created, 60 mm away from the front plane, in the direction of the extrusion of sketch 2. The entire extruded body was then mirrored in relation to this new plane 1, as shown in Figure C3.

Figure C3: New plane 1 and mirroring of the crankcase.

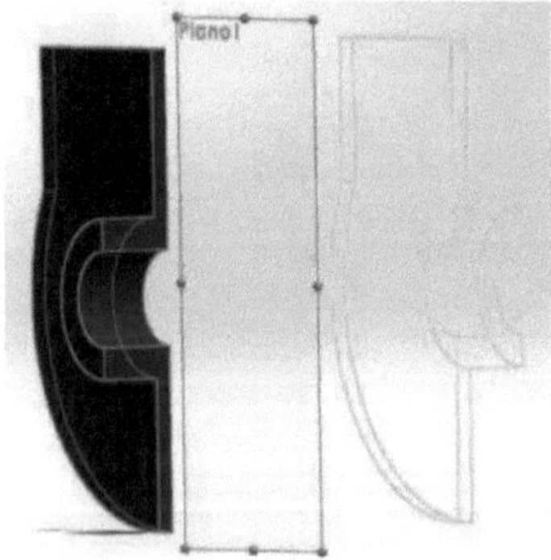

Source: Author.

From one of the inner faces, sketch 3 was created to form the side wall of the carter. By extruding this sketch to the other face, as shown in Figure C4, the desired wall was formed.

Figure C4: a - sketch 3; b - formation of the side wall.

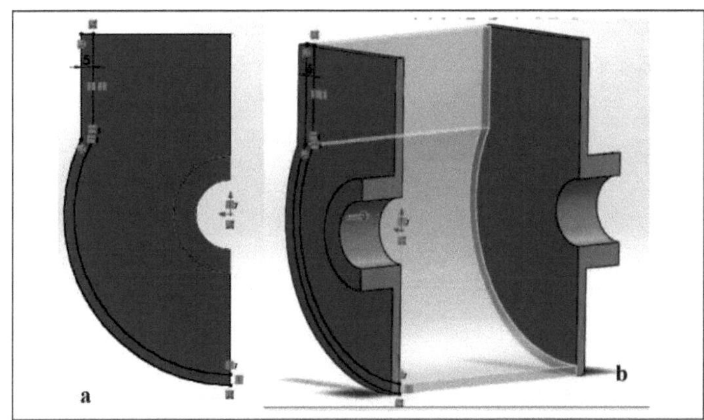

Source: Author.

The entire body *already* created was mirrored in relation to the right plane in order to finalize the piece. Details such as the type of material chosen, forged aluminum, and the finished piece can be seen in Figure C5.

Figure C5: Carter finished.

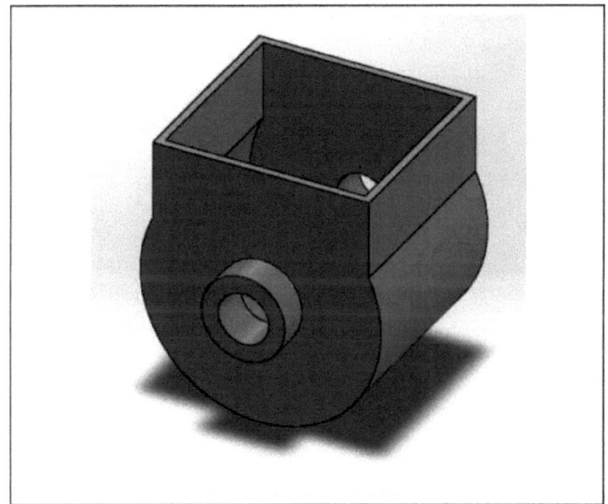

Source: Author.

APPENDIX D - Crankshaft gear 1

Sketch 1 was created in the right-hand plane, where a 60 mm diameter circumference was made and a 15 mm extrusion was made in the middle plane (FIG. D1).

Figure D1: Sketch 1 and extrusion of crankshaft gear 1.

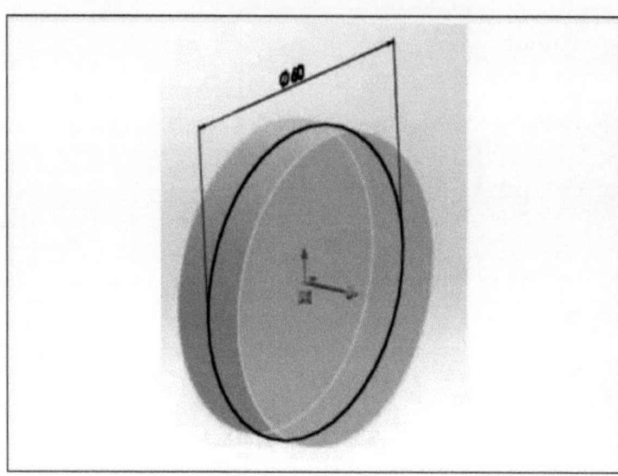

Source: Author.

A sketch 2, shown in Figure D2, was created on one of the faces of the circumference and then extruded to the opposite face, obtaining the first tooth of the gear. Using the circular pattern command, 16 teeth were created on the gear with equal spacing between them.

Figure D2: a - sketch 2; b - teeth of crankshaft gear 1.

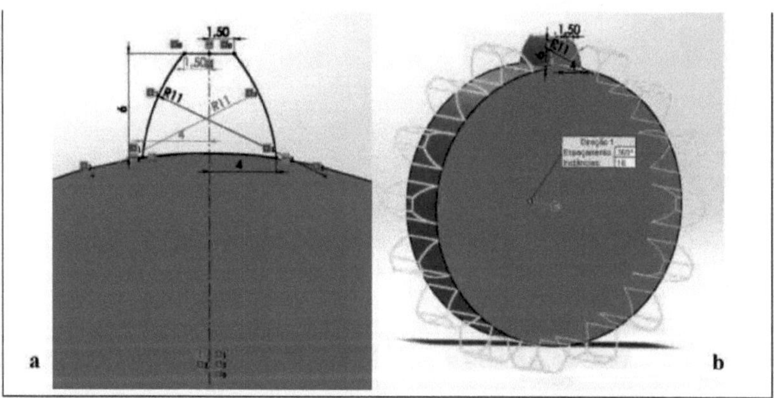

Source: Author.

From the front plane, sketch 3 was created and a 45 mm circumference was made, which was extruded through the middle plane by 36 mm, as shown in Figure D3.

Figure D3 : Sketch 3 and extrusion of gear 1 from the crankshaft.

Source: Author.

On one of the faces extruded by sketch 3, sketch 4 was created, shown in Figure F4. Next, an extruded cut with a through condition was made to form the hole through which the crankshaft shaft will be attached.

Figure D4: a - sketch 4; b - extruded section of crankshaft gear 1.

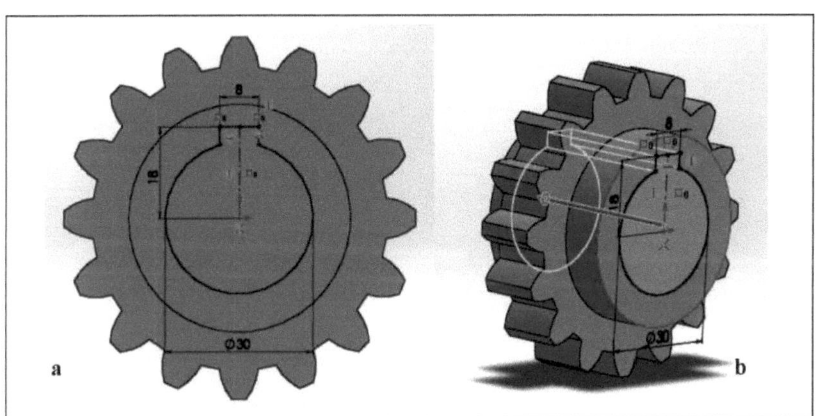

Source: Author.

Figure D5 shows the completed gear 1 of the crankshaft and the cast carbon steel chosen as the material type.

Figure D5: Finished crankshaft gear 1.

Source: Author.

APPENDIX E - Crankshaft gear 2

The process for creating crankshaft gear 2 was similar to that for gear 1. From the right-hand plane, sketch 1 was created and a circle 83 mm in diameter was made with an extrusion of 15 mm from the middle plane, as shown in Figure E1.

Figure E1: Sketch 1 and extrusion of gear 2 from the crankshaft.

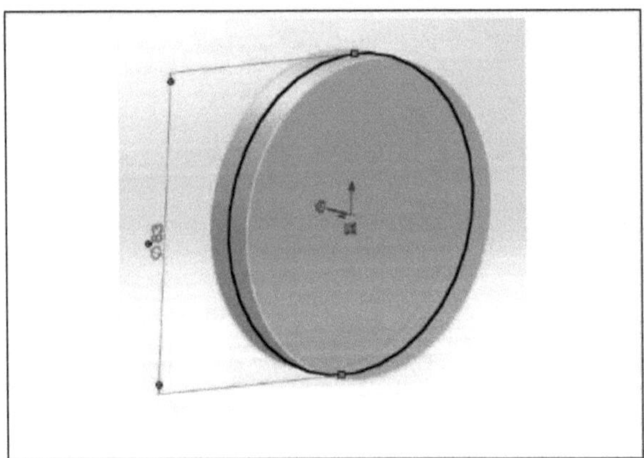

Source: Author.

Sketch 2, shown in Figure E2, was created on one of the previously extruded faces of sketch 1. An extrusion was made from one face to the other, forming a tooth and the circular pattern feature was used, with equal spacing, to create the other teeth of the gear.

Figure E2: Sketch 2 and teeth of gear 2 on the crankshaft.

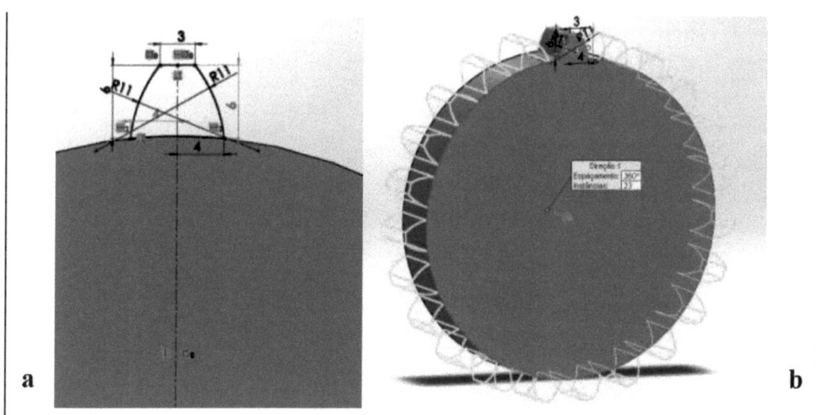

Source: Author.

Sketch 3 was opened from the right plane and a 45 mm diameter circle was created in the center of the gear. As shown in Figure E3, a 36 mm extrusion was made in the middle plane.

Figure E3: Sketch 3 and extrusion of gear 2 from the crankshaft.

Source: Author.

In one of the previously created faces of sketch 3, sketch 4 was created, shown in Figure E4. An extrusion was made with the through condition to form the hole where the crankshaft will be attached.

Figure E4: a - sketch 4; b - extruded section of crankshaft gear 2.

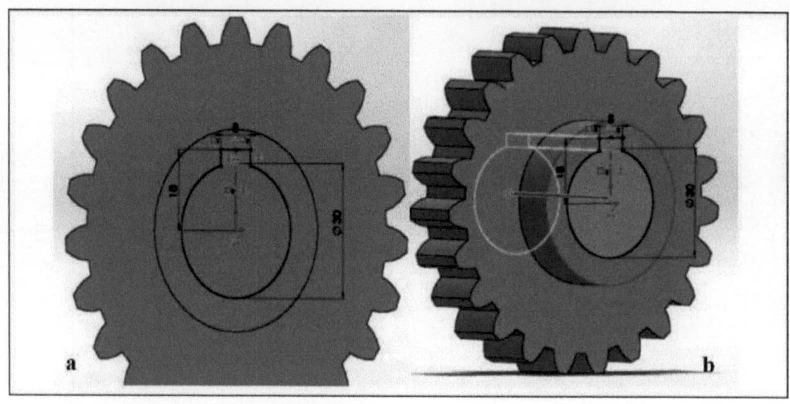

Source: Author.

Figure E5 shows gear 2 of the crankshaft finished with the type of material chosen being cast carbon steel.

Figure E5: Gear 2 of the crankshaft finished.

Source: Author.

APPENDIX F - Piston camshaft gearing

Sketch 1 was opened from the right plane, creating a 166 mm circumference, as shown in Figure F1. A 15 mm extrusion was then made through the middle plane of this circumference to form the gear body.

Figure F1: Sketch 1 and extrusion of the piston camshaft gear.

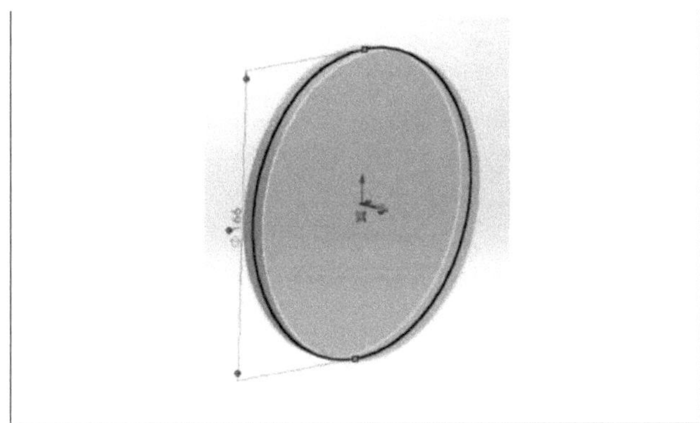

Source: Author.

From one of the previously extruded faces, sketch 2 was opened, where the gear tooth was drawn, as seen in Figure F2. The tooth outline was extruded to the other face of the gear and replicated with the circular pattern command, with equal spacing between them.

Figure F2: a - sketch 2; b - gear teeth on the piston camshaft.

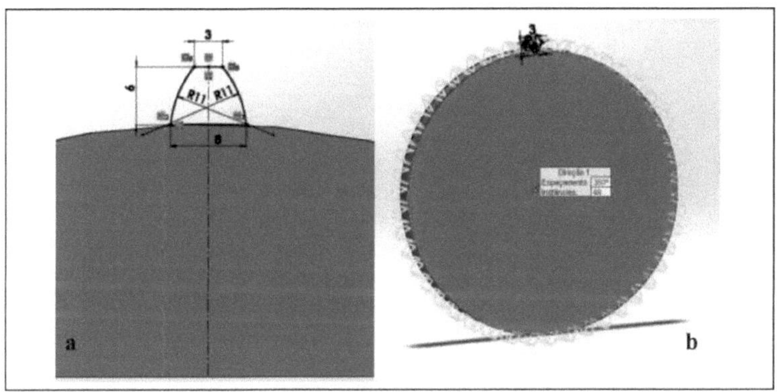

Source: Author.

Sketch 3 was created in the direct plane and then a 90 mm diameter circle was drawn in the center of the gear. It was then extruded through the middle plane by 40 mm, as can be seen in Figure F3.

Figure F3: Sketch 3 and central extrusion of the piston camshaft gear.

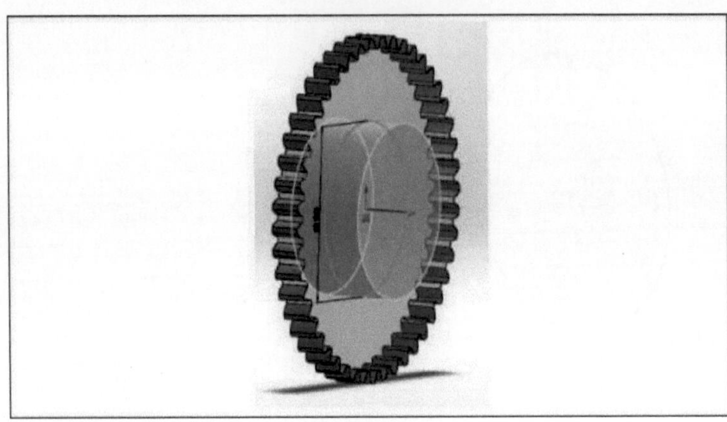

Source: Author.

From one of the extruded faces of sketch 3, sketch 4 was opened, shown in Figure F4. An extruded cut was made, with the through condition, to form the hole that will be coupled to the control shaft of the clutch.

Figure F4: Sketch 4 and extruded section of the piston camshaft gear.

Source: Author.

To create the holes around the gear, sketch 5 was opened from the right-hand plane and a 128 mm construction circle was drawn. On top of this construction circle, another 13 mm diameter circle was made, as shown in Figure F5. Next, a 15 mm extruded cut was made through the middle plane and, using the circular pattern command with 8 instances and equal spacing between them, the holes were obtained.

Figure F5: a - sketch 5; b - gear orifices on the piston camshaft.

Source: Author.

The cast carbon steel chosen as the material and the finished part can be seen in Figure F6.

Figure F6: Completed piston camshaft gear.

Source: Author.

APPENDIX G - Camshaft gearing

Sketch 1 was created from the right-hand plane, where a 120 mm circumference was extruded by 15 mm from the middle plane, as shown in Figure G1.

Figure G1: Sketch 1 and extrusion of the camshaft gear.

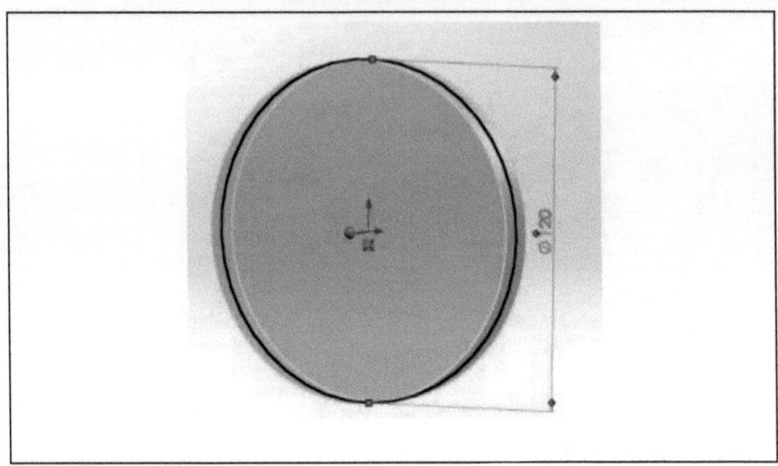

Source: Author.

Starting with sketch 2, the gear tooth was made on one of the faces and then extruded onto the other face. With the circular pattern feature with 32 instances and equal spacing, the others were obtained, as seen in Figure G2.

Figure G2: a - sketch 2; b - camshaft gear teeth.

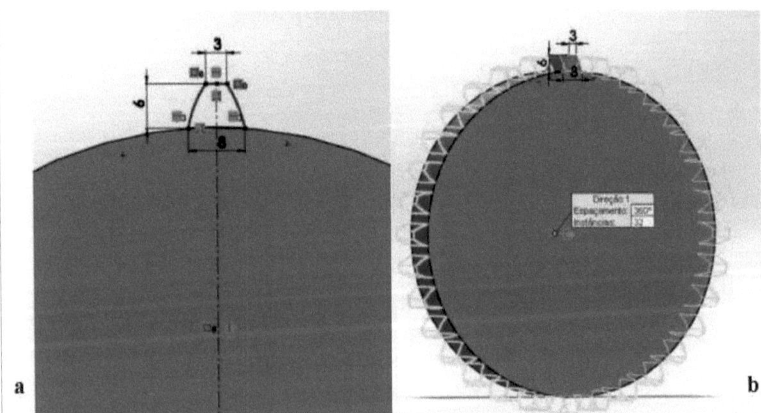

Source: Author.

In the right-hand plane, sketch 3 was created, where a 40 mm diameter circle was made in the center of the gear and extruded by 36 mm through the middle plane, as shown in Figure G3.

Figure G3 : Sketch 3 and extrusion of the camshaft gear.

Source: Author.

In one of the faces of sketch 3, sketch 4 was opened in order to create the holes through which the crankshaft shaft would pass. This new sketch was extruded through a through condition, as can be seen in Figure G4.

Figure G4: a - sketch 4; b - extruded section of the camshaft gear.

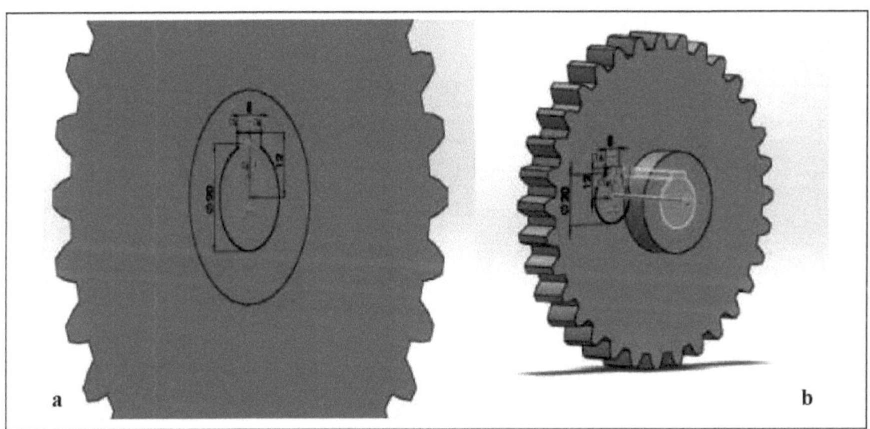

Source: Author.

Sketch 5 was created from the right plane, where a 78.5 mm diameter circle was drawn from the center of the gear. Another circle with a diameter of 13 mm was drawn on the contour of this circle. A 15 mm extruded cut was made through the middle plane, forming a hole in the gear body. Using the circular pattern, with equal spacing and 7 instances, the other holes were obtained. These details can be seen in Figure G5.

Figure G5: a - sketch 5; b - camshaft gear holes.

Source: Author.

To finalize the construction of this part, cast carbon steel was chosen as the material, as can be seen in Figure G6.

Figure G6: Completed camshaft gear.

Source: Author.

APPENDIX H - Connecting rod

The creation of the connecting rod began in the front plane, where sketch 1 was made, followed by a 20 mm extrusion in the middle plane, as shown in Figure H1.

Figure H1: a - sketch 1; b - connecting rod extrusion.

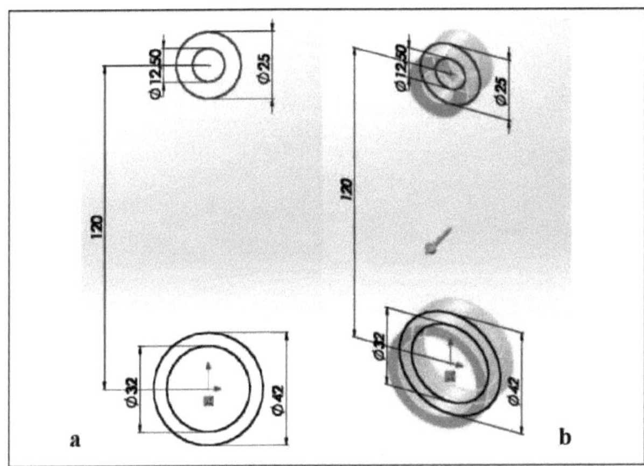

Source: Author.

Again from the front plane, sketch 2 was created and extruded by 12 mm from the middle plane to form the connecting rod body (FIG. H2).

Figure H2: a - sketch 2; b - connecting rod body extrusion.

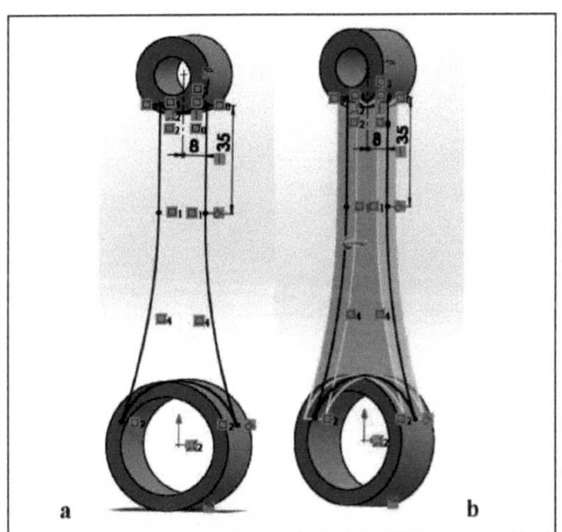

Source: Author.

The end counterweights were created from the front plane, using sketch 3, and extruded by 12 mm from the middle plane (FIG. H3).

Figure H3: a - sketch 3; b - extrusion of the connecting rod counterweights.

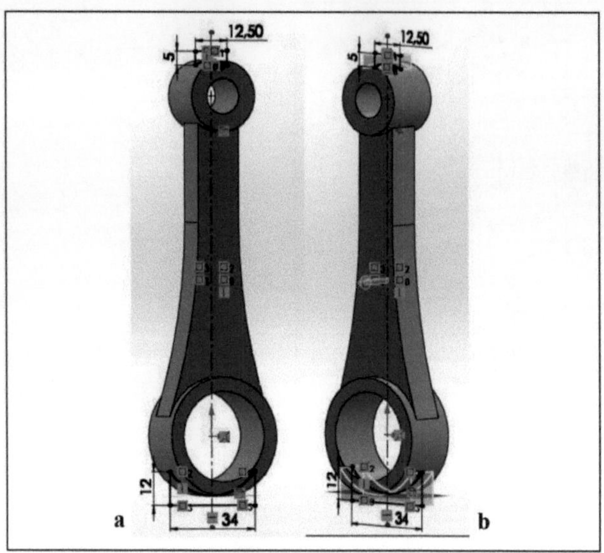

Source: Author.

To finish the part, a 1.5 mm fillet was used on the edges joining the upper and lower holes to the body and counterweights. Forged steel was then chosen as the material type (FIG. H4).

Figure H4: Finished connecting rod.

Source: Author.

APPENDIX I - Piston balancer

The creation of the rocker started from the right-hand side, opening sketch 1 and drawing the entire body of the piece, as shown in Figure I1.

Figure I1: Sketch 1 of the piston rocker arm.

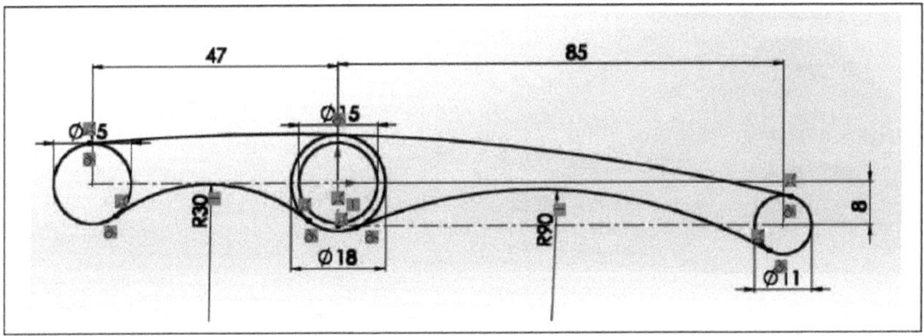

Source: Author.

The rocker body was extruded in three stages. First, the two end circumferences were extruded by 15 mm in the middle plane. Next, the two central circumferences were extruded through the mid-plane by 20 mm, forming a hollow cylinder. Finally, the rest of the body was extruded through the mid-plane by 5 mm. The extrusions are shown in Figure I2.

Figure I2: Construction stages of the rocker arm: a - sketch drawing; b - extruding the ends of the central hole; c - extruding the body.

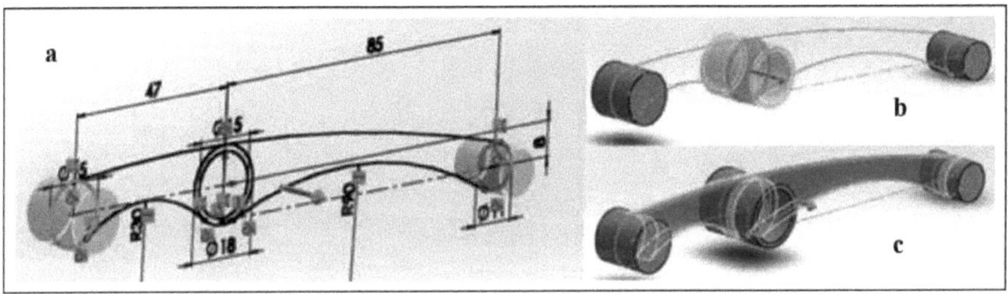

Source: Author.

For a final finish, fillets with a radius of 2 mm were applied to the joints of the rocker body, as shown in Figure I3. The type of material chosen was carbon steel.

Figure I3: Finished piston rocker arm.

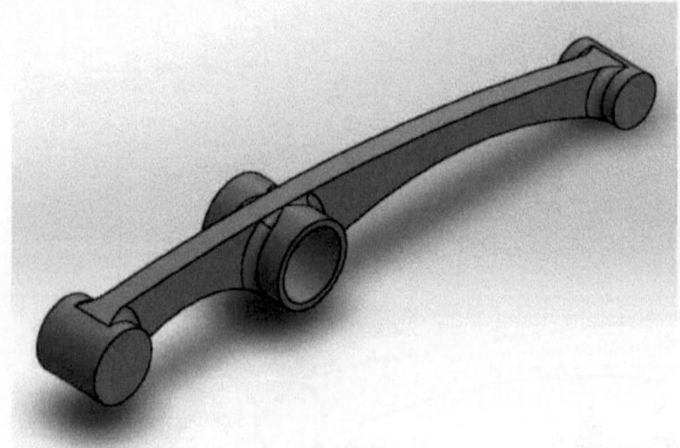

Source: Author.

APPENDIX J - Valve rocker arm

The rocker arm that moves the valves was created in the same way as the piston rocker arm, but with different measurements, as shown in Figure J1.

Figure J1: Sketch 1 of the valve rocker arm.

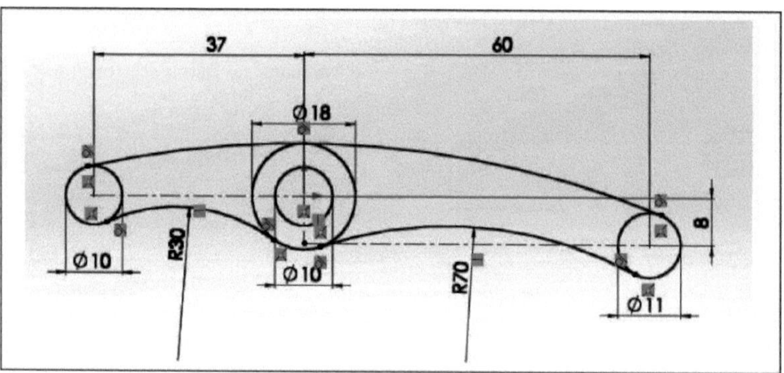

Source: Author.

The circumferences at the ends were extruded 8 mm apart in the middle plane. In the center, a 20 mm mid-plane extrusion was made between the two circumferences, forming a hollow cylinder. Finally, a 5 mm extrusion was made in the rest of the rocker body, as shown in Figure J2.

Figure J2: Construction stages of the rocker arm: a- sketch drawing;

b - extrusion of the ends of the central hole; c - extrusion of the body.

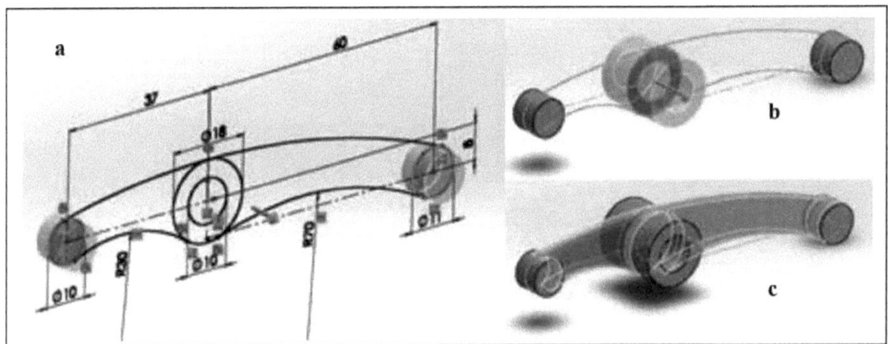

Source: Author.

Finally, 2 mm fillets were applied to the rocker joints and carbon steel was chosen as the material type (FIG. J3).

Figure J3: Valve rocker arm completed.

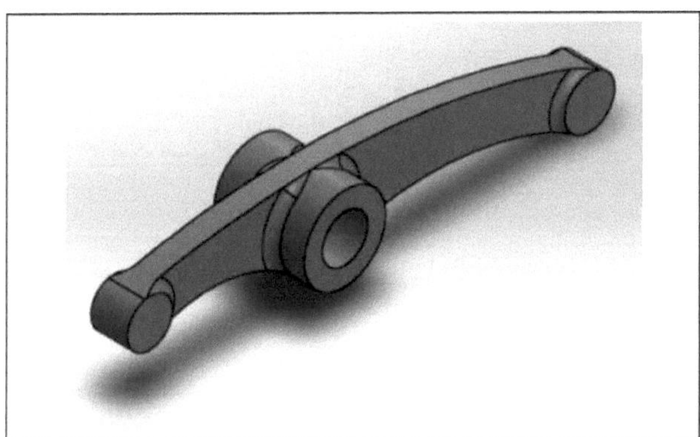

Source: Author.

APPENDIX K - Variable compression piston rod

For the piston rod, sketch 1 was created and, on it, a circumference of 25 mm was created, starting from the origin, where it was extruded in 50 mm, as shown in Figure K1.

Figure K1: Sketch 1 and extrusion of the rod base.

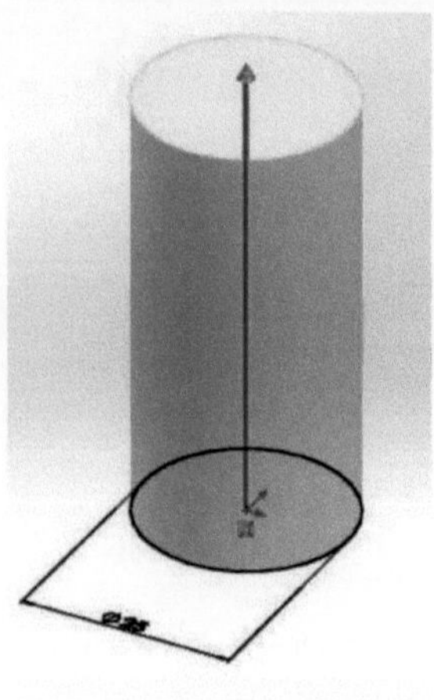

Source: Author.

On the face of the previously extruded circumference, sketch 2 was opened, where a 12 mm circumference was made from the origin. A blind extrusion of 168.33 mm was then made, as can be seen in Figure K2.

Figure K2: Sketch 2 and rod extrusion.

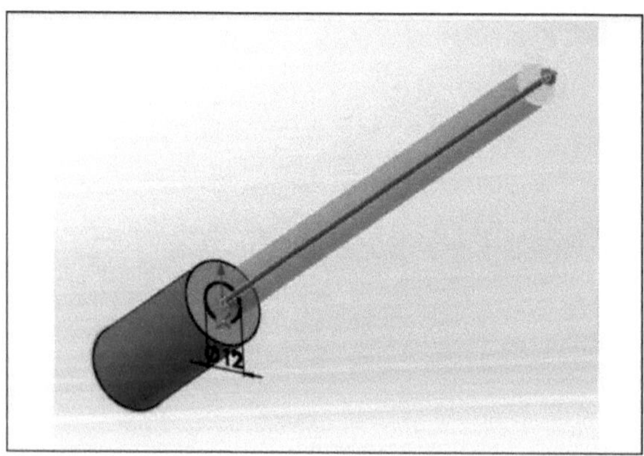

Source: Author.

Sketch 3 was created at the end of the circumference of sketch 2, and on it a 20 mm circumference from the origin. Next, a 3 mm blind extrusion was made and a 4 mm dome was applied to the upper face of the circle (FIG. K3).

Figure K3: a - sketch 3; b - piston rod extrusion and dome.

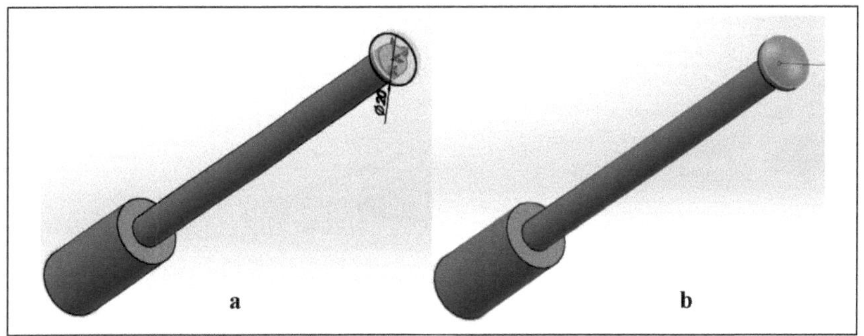

Source: Author.

After applying the dome, a 10 mm radius fillet was applied at the junction of sketch 1 and sketch 2. Then another fillet with a radius of 5 mm was applied to the base of the rod, as shown in Figure K4.

Figure K4: Sketch of the rod base fillets: a - top; b - bottom.

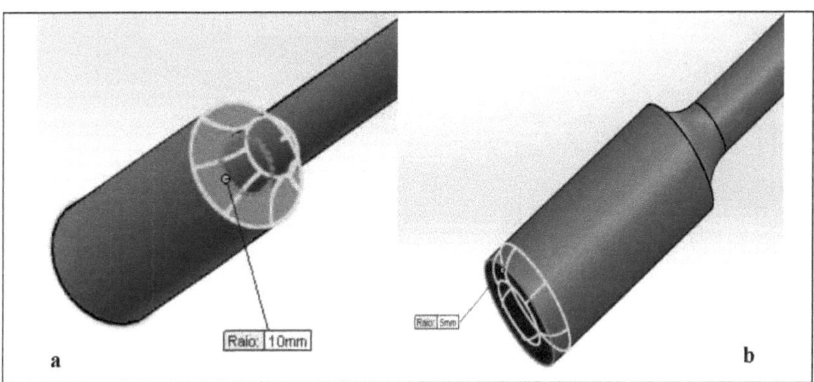

Source: Author.

The final piece, as well as the type of material chosen for the finish, polished steel, can be seen in Figure K5.

Figure K5: Finished piston rod.

Source: Author.

APPENDIX L - Valve rods

From the upper plane, sketch 1 was created and a 20 mm circle centered on the origin was made. This circle was then blind extruded by 45 mm, as shown in Figure L1.

Figure L1: Sketch 1 and extrusion of the rod base.

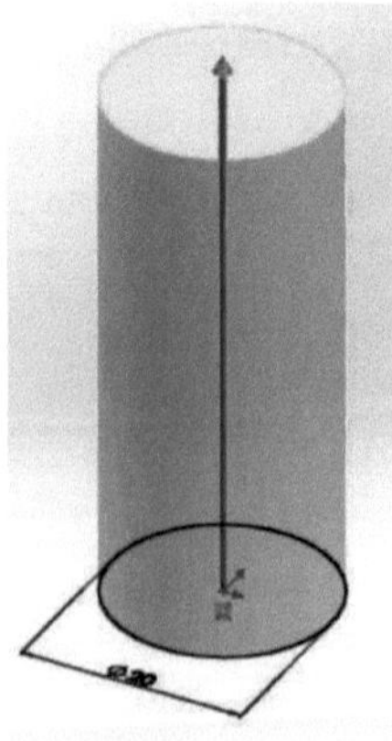

Source: Author.

Sketch 2 was created on the upper face of the previously extruded circumference, and then an 8 mm diameter circumference was made from the origin. After this circumference, a 207.5 mm blind extrusion was made, as can be seen in Figure L2.

Figure L2: Sketch 2 and rod extrusion.

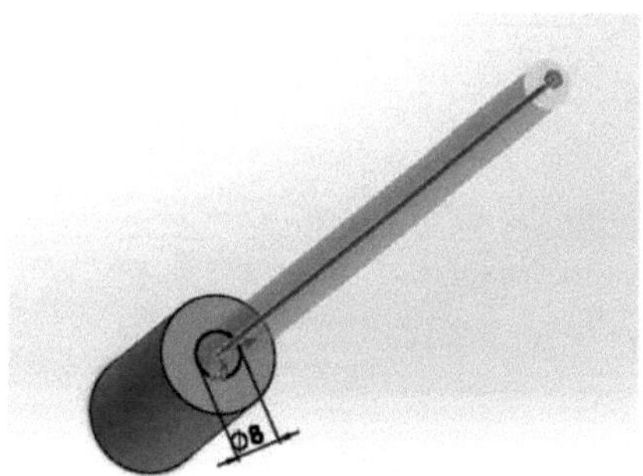

Source: Author.

On top of the 207.5 mm extrusion made from sketch 2, sketch 3 was opened. In this new sketch, a circumference of 12 mm was made from the origin, which was extruded blindly by 2 mm. To finalize this sketch, a 3 mm dome was applied to the extruded face, as can be seen in Figure L3.

Figure L3: a - sketch 3; b - extrusion of the valve rod dome.

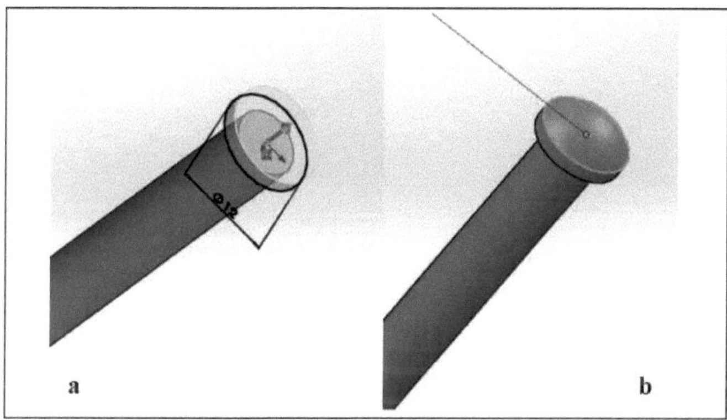

Source: Author.

For the final finish, a 10 mm fillet was applied at the junction of the larger circumference of sketch 1 and the smaller one of sketch 2. Another 5 mm fillet was then applied to the lower end of the rod, as shown in Figure L4.

Figure L4: Valve rod fillets.

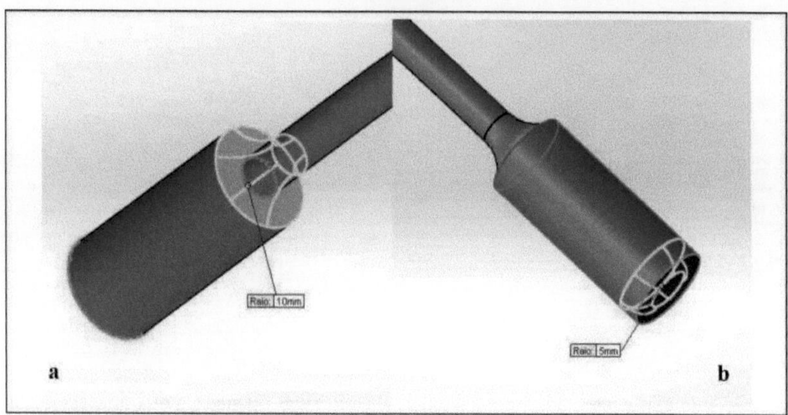

Source: Author.

Figure L5 shows the final result and the type of material chosen for the piece, which was polished steel.

Figure L5: Finished valve rod.

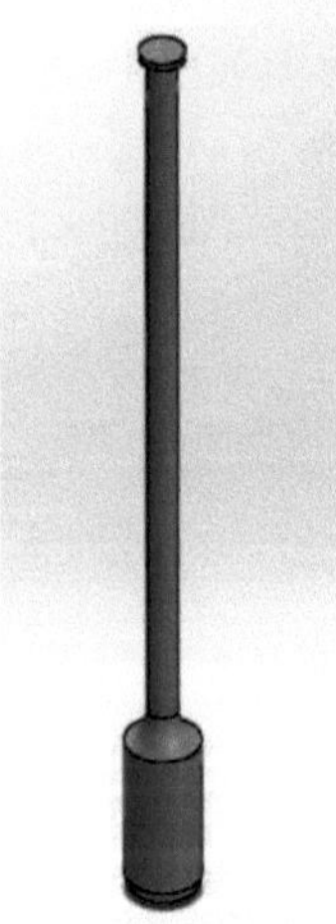

Source: Author.

APPENDIX M - Valves

The valve was built using only a sketch created from the frontal plan. This sketch, together with the detail of the valve head, can be seen in Figure M1, which represents half of the valve design.

Figure M1: a - sketch 1 of the valve; b - enlargement of image M1a.

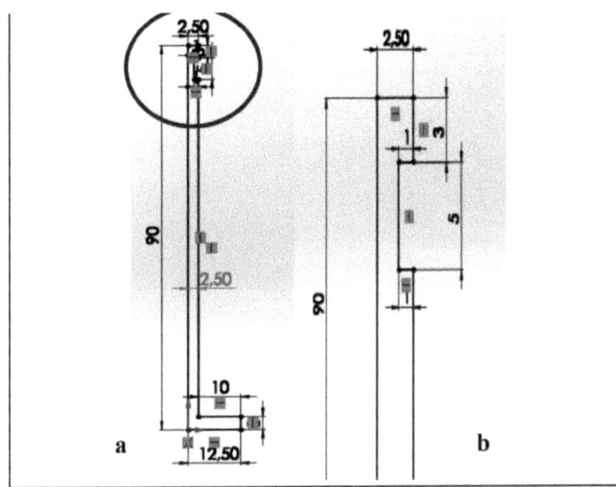

Source: Author.

Once sketch 1 had been made, we applied the rebound/revolutionized base feature, making it blind and choosing the 90 mm line as the axis of revolution. Figure M2 shows the entire valve body.

Figure M2: Valve rotation.

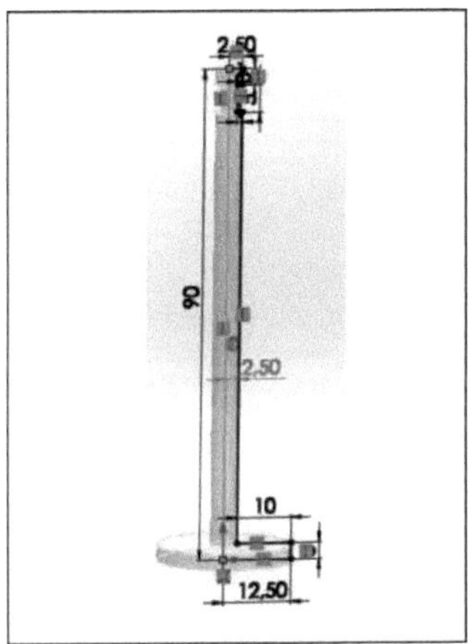

Source: Author.

For a final finish, a 6 mm radius fillet was applied to the junction between the base and the valve body, as shown in Figure M3.

Figure M3 : Valve fillet.

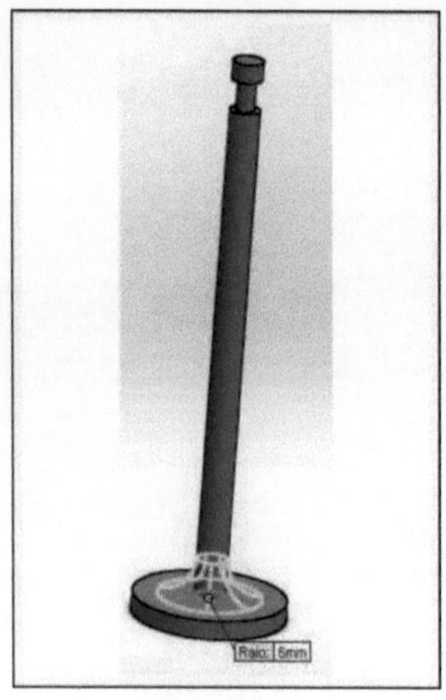

Source: Author.

Figure M4 shows the finished valve and the type of material chosen, machined steel.

Figure M4: Finished valve.

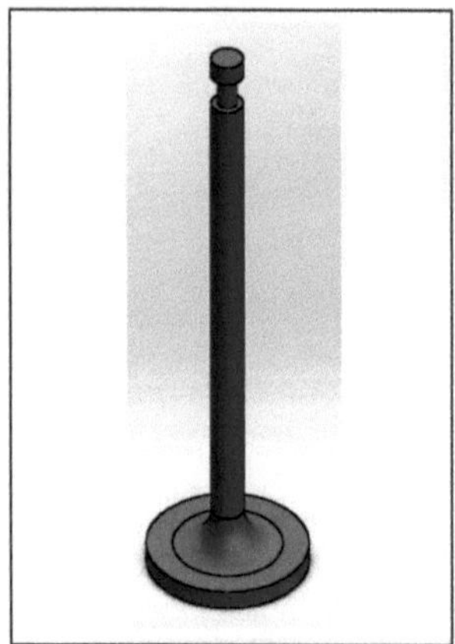

Source: Author.

APPENDIX N - Crankshaft

Sketch 1 started at the front, with the construction of a 30 mm circumference and an 88 mm blind extrusion, as shown in Figure N1.

Figure N1: Sketch 1 and crankshaft extrusion.

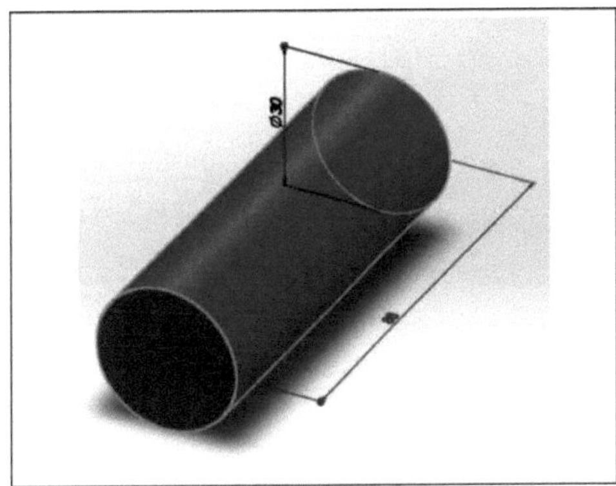

Source: Author.

For the construction of the crankshaft counterweight, sketch 2 was created from one of the front faces of sketch 1. A 12 mm blind extrusion was then made, as can be seen in Figure N2. As you can see, the drawing of the cam profile is only an approximation of a real model.

Figure N2: a - sketch 2; b - counterweight extrusion.

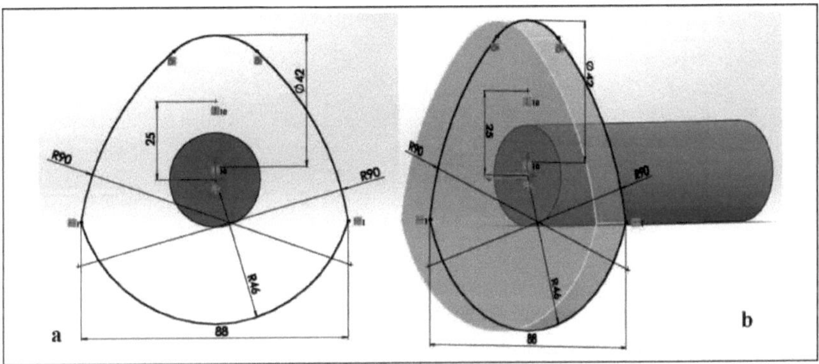

Source: Author.

Figure N3 shows sketch 3, where a 32 mm circumference was opened at a distance of 25 mm from the origin. Outline 3 was then extruded 15 mm outwards to form half of the crankshaft wedge.

Figure N3: a - sketch 3; b - construction of the crankshaft wedge.

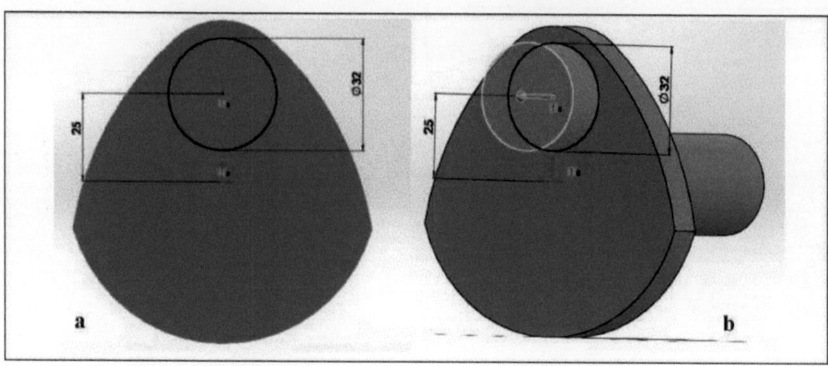

Source: Author.

To build the key, which will connect the shaft to the gears, we selected the end of sketch 1 and created sketch 4. We then extruded 36 mm towards the counterweights, as shown in (FIG. N4).

Figure N4: a - sketch 4; b - crankshaft keyway extrusion.

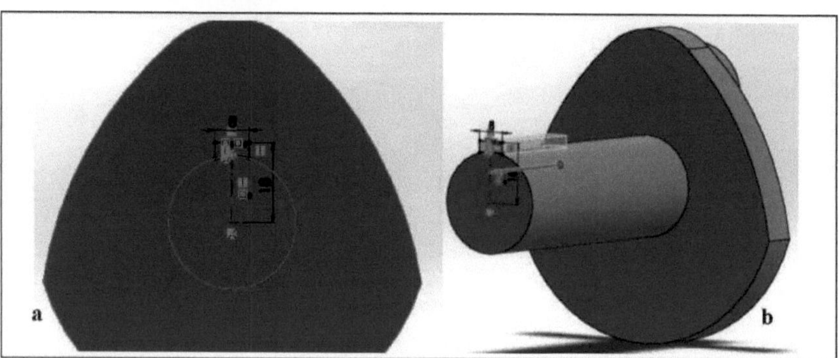

Source: Author.

After extruding sketch 4, half of the crankshaft was ready. The entire body already built was then mirrored in relation to the end of sketch 3, forming the final part. Finally, as can be seen in Figure N5, forged steel was chosen as the material for the part.

Figure N5: Finished crankshaft.

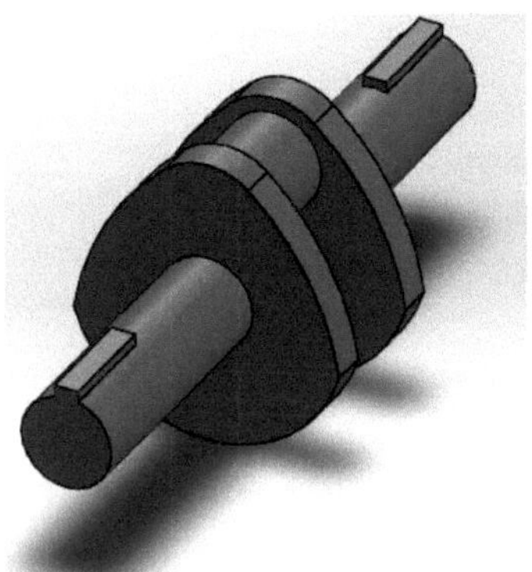

Source: Author.

APPENDIX O - Piston rocker arms

The embolus rocker arm was created from the right-hand plane. Opening sketch 1 and making a circumference of 10 mm from the origin, a blind extrusion of 98.5 mm was made (FIG. O1).

Figure O1: Sketch 1 and extrusion of the plunger rocker bearing.

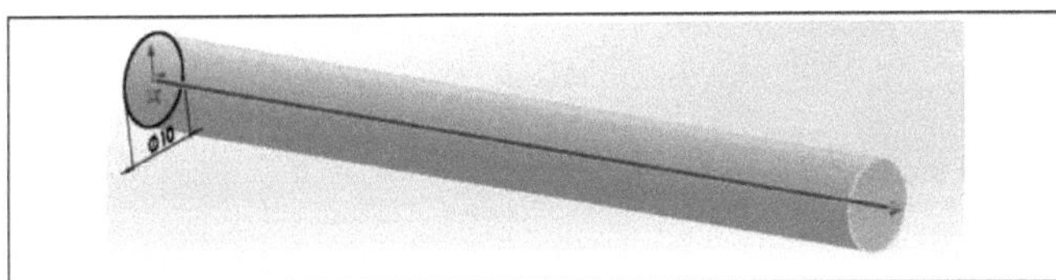

Source: Author.

The offset on the shaft, where the rocker arm will be, was started by opening a sketch 2 on the face of the previously extruded circumference. A 15 mm circumference was created from the origin, which was extruded blind by 5 mm (FIG. O2).

Figure O2: Sketch 2 and extrusion of the rocker arm bearing shoulder.

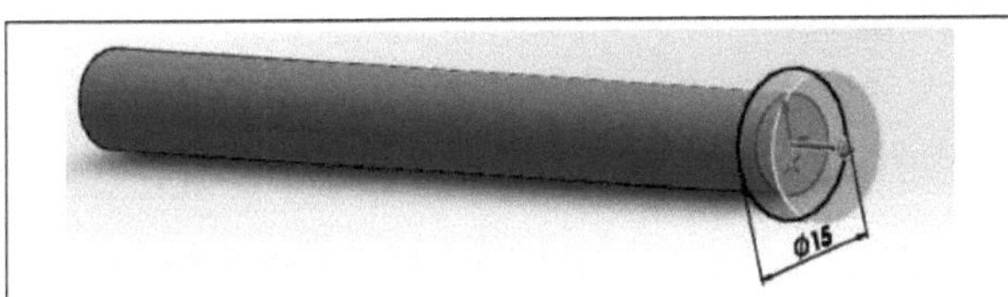

Source: Author.

Sketch 3 was created from the extruded face, followed by a 10 mm circle centered on the origin. Then a 20 mm blind extrusion was made (FIG. O3).

Figure O3: Sketch 3 and extrusion of the rocker arm bearing.

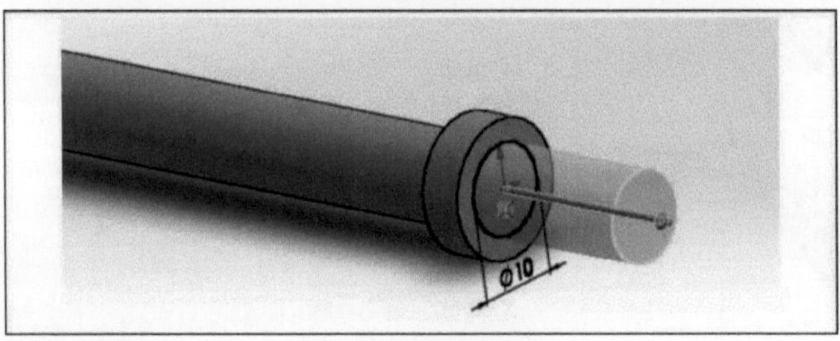

Source: Author.

To finalize the rebound, sketch 4 was opened on the face of the extruded circumference of sketch 3. As can be seen in Figure O4, another 15mm circle centered on the origin was opened, this one extruded 5mm blind.

Figure O4: Sketch 4 and extrusion of the rocker arm shoulder.

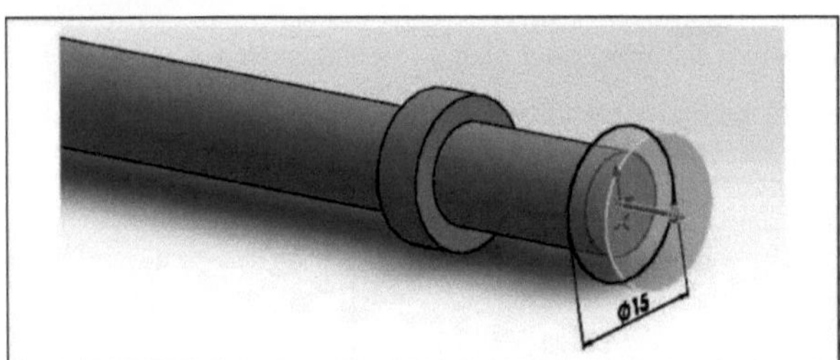

Source: Author.

From the extruded face of sketch 4, sketch 5 was opened and a 10 mm circle centered on the origin was made. A 100 mm blind extrusion was made on this circumference (FIG. O5).

Figure O5: Sketch 5 and extrusion of the rocker arm bearing.

Source: Author.

For the final finish, it can be seen in Figure O6 that the type of material chosen was matt steel.

Figure O6: Completed piston rocker arm bearing.

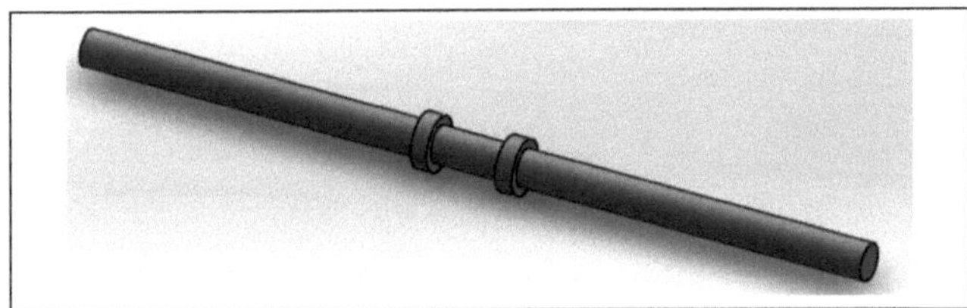

Source: Author.

APPENDIX P - Valve rocker arm bearings

Sketch 1 was created from the right-hand plane, followed by a 10 mm circle from the origin. It was then extruded 83 mm blind, as can be seen in Figure P1.

Figure P1: Sketch 1 and extrusion of the valve rocker arm.

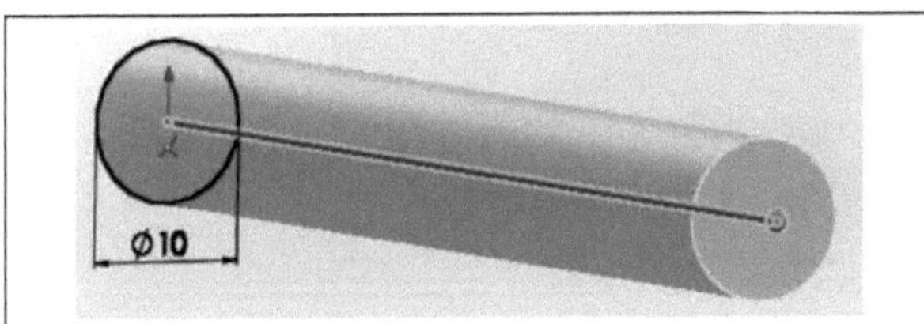

Source: Author.

Sketch 2 was created from the extruded face of sketch 1 and then a 14 mm circle centered on the origin. Then a 64 mm blind extrusion was made, as shown in Figure P2.

Figure P2: Sketch 2 and extrusion of the valve rocker arm.

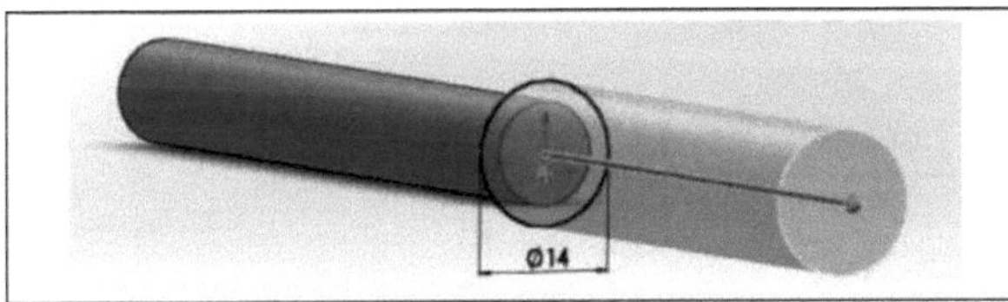

Source: Author.

Sketch 3 was made on the face of the previously extruded circumference and on it a circumference of 10 mm, starting from the origin. Then a blind extrusion of 81.5 mm was made, as shown in Figure P3.

Figure P3: Sketch 3 and valve rocker arm bearing extrusion

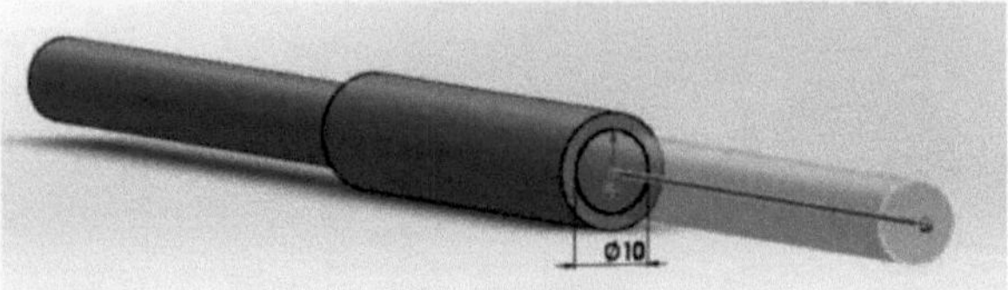

Source: Author.

It can be seen in Figure P4 that matt steel was chosen as the type of material for the final finish.

Figure P4: Completed valve rocker arm bearing.

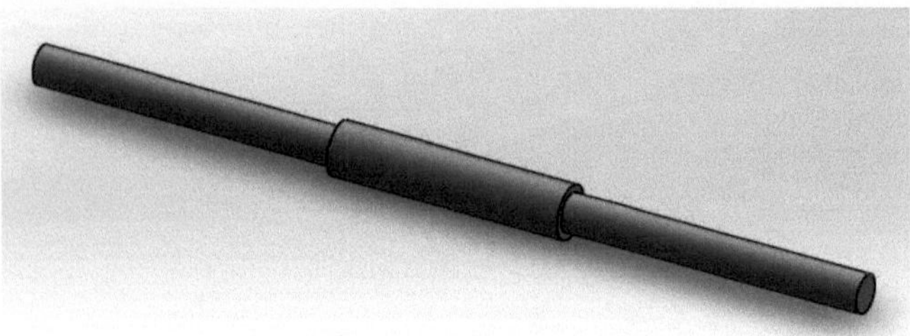

Source: Author.

APPENDIX Q - Pin

Construction of the pin began on the right-hand side, with sketch 1, where two circumferences were made. The larger one, measuring 12.5 mm, and the smaller one, measuring 7.5 mm, were blind extruded by 50 mm to form the hollow cylinder (FIG. Q1).

Figure Q1: Sketch 1 and pin extrusion.

Source: Author.

The final finish was given by applying a 0.5 mm fillet at both ends and cast aluminium was chosen as the type of material (FIG. Q2).

Figure Q2: Finished pin.

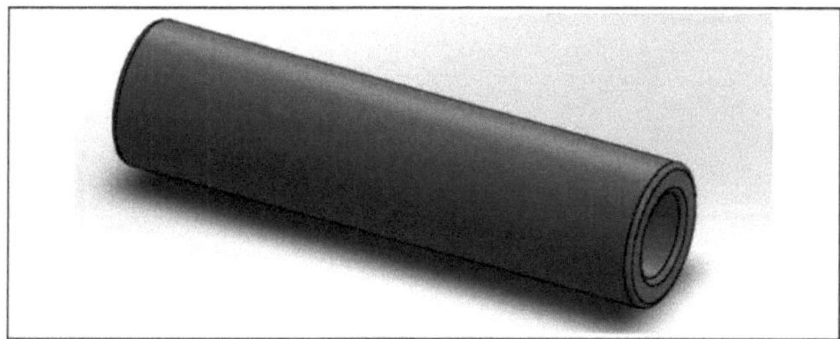

Source: Author.

APPENDIX R - Variable compression piston

Like the valve, the embolus was also made with just one sketch created from the right-hand plane, as can be seen in Figure R1.

Figure R1: a - sketch 1 of the piston; b - enlargement of Figure R1a.

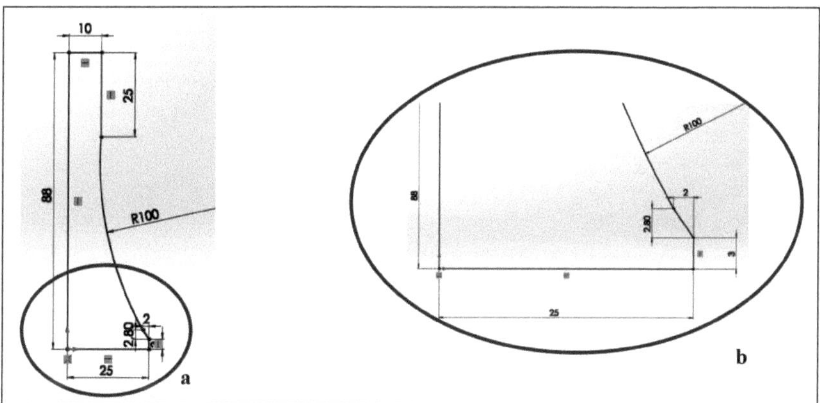

Figure Q2: Finished pin.

To form the entire piston, a blind revolution was made in sketch 1, with the 88 mm line as its central axis, as shown in Figure R2. Finally, a 2 mm radius fillet was applied to the head of the piston and carbon steel was chosen as the type of material.

Figure R2: Reproduction of the symbol: a - revolution of the symbol; b - finished symbol.

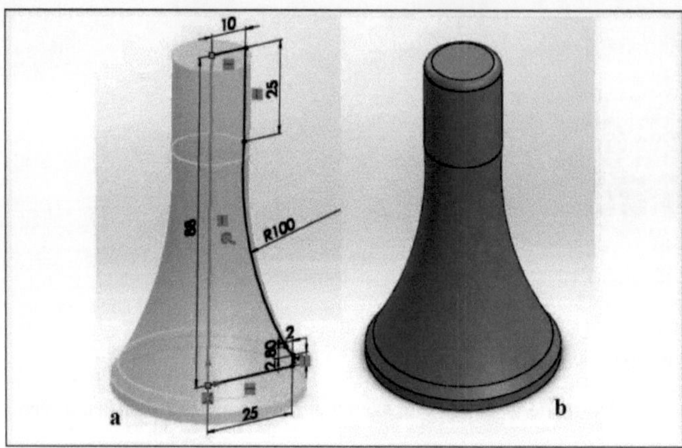

Source: Author.

APPENDIX S - Piston cam shaft

Sketch 1 was created from the right-hand plane, where a circle was made from the origin with a diameter of 60 mm. This circle was then extruded into 105 mm as shown in Figure S1.

Figure S1: Sketch 1 and extrusion of the piston cam shaft.

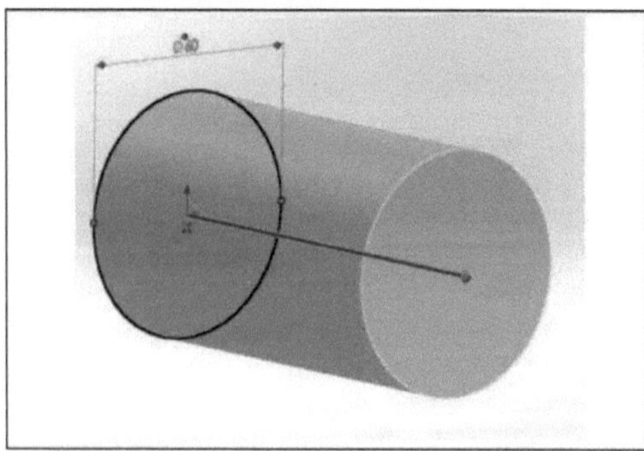

Source: Author.

On the face of the previously extruded circumference, a new cam outline 2 was opened and a 20 mm blind extrusion was made (FIG. S2).

Figure S2: a - sketch 2; b - cam extrusion.

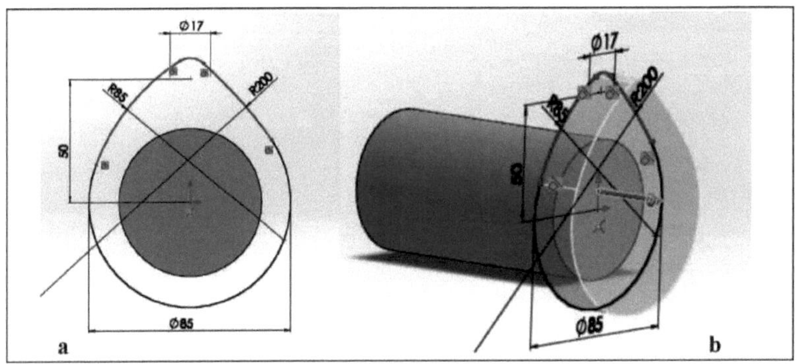

Source: Author.

Sketch 3 was created on the extruded face of sketch 2, and a 60 mm diameter circle was made from the origin. A 103.5 mm extrusion was then made, as shown in Figure S3.

Figure S3: Extrusion of sketch 3 of the piston camshaft.

Source: Author.

To create the key that will connect the shaft to the gear, a blank 4 was opened on the extruded face of blank 3, as shown in Figure S4. This blank 4 was then extruded 40 mm blind onto the cylinder that forms the shaft.

Figure S4: a - sketch 4; b - keyway extrusion.

Source: Author.

When finalizing the piston cam shaft, machined steel was chosen as the type of material (FIG. S5).

Figure S5: Completed piston cam shaft.

Source: Author.

APPENDIX T - Valve camshaft

Sketch 1 was opened from the right-hand plane, where a 20 mm diameter circle was constructed from the origin. This was blind extruded by 65.5 mm, as shown in Figure T1.

Figure T1: Sketch 1 and valve cam shaft extrusion.

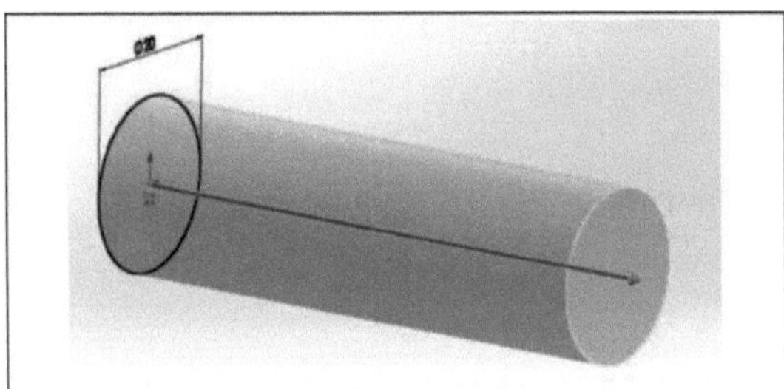

Source: Author.

At the end of the previously extruded face, outline 2 of the cam was created, as shown in Figure T2. This outline was blindly extruded 15 mm outwards to form one of the shaft's cams.

Figure T2: a - sketch 2; b - cam extrusion.

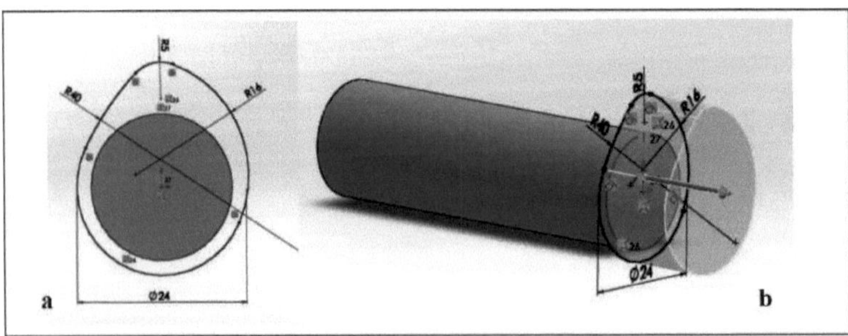

Source: Author.

Selecting the extruded cam face, a new sketch 3 was opened where a 20 mm diameter circumference was created which was extruded 69 mm from the origin (FIG. T3).

Figure T3: Extrusion of sketch 3 of the valve camshaft

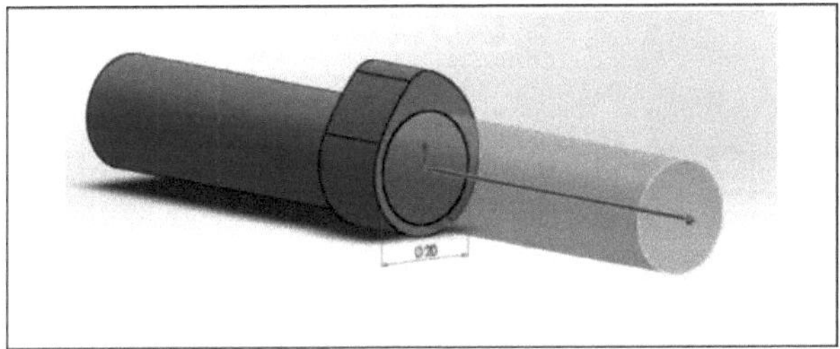

Source: Author.

On the extruded face of sketch 3, sketch 4 of the second cam was created, which is identical to sketch 2 of the first cam. However, it was rotated 270 degrees relative to its origin and then extruded 15 mm to form the second cam (FIG. T4).

Figure T4: a - sketch 4; b - cam extrusion.

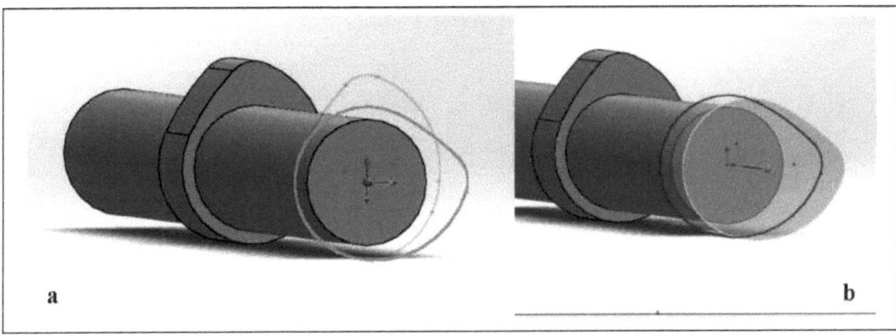

Source: Author.

On the outer face of the second cam, sketch 5 was created, and on it a 20 mm circumference from the origin that was extruded in 64 mm (FIG. T5).

Figure T5: Extrusion of sketch 5 from the valve camshaft.

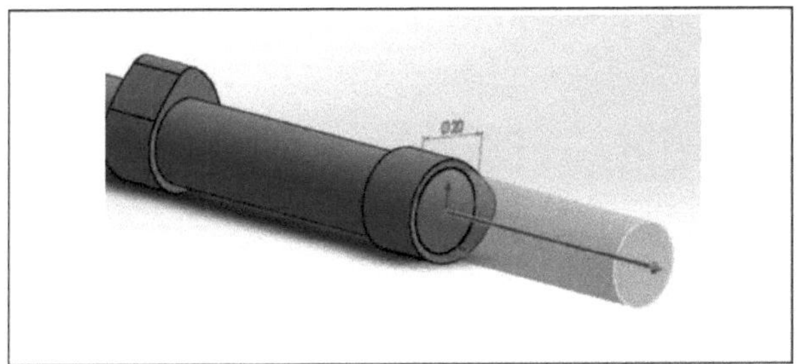

Source: Author.

The shaft key was made by blindly extruding 36 mm of blank 6, open on the extruded face of blank 5, as shown in Figure T6.

Figure T6: a - sketch 6; b - keyway extrusion.

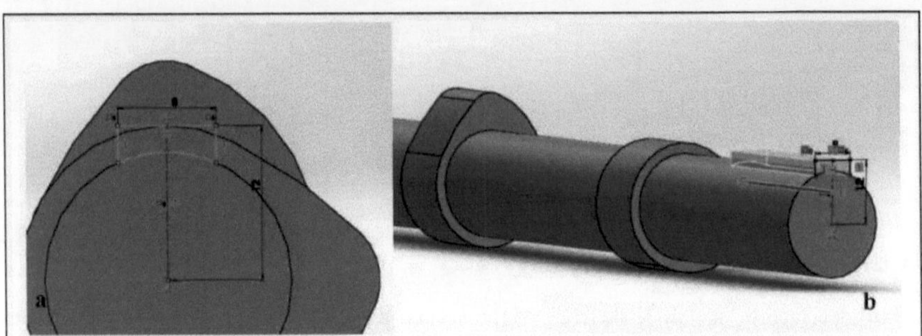

Source: Author.

For the finished valve cam shaft, machined steel was chosen as the material type (FIG. T7).

Figure T7: Completed valve camshaft.

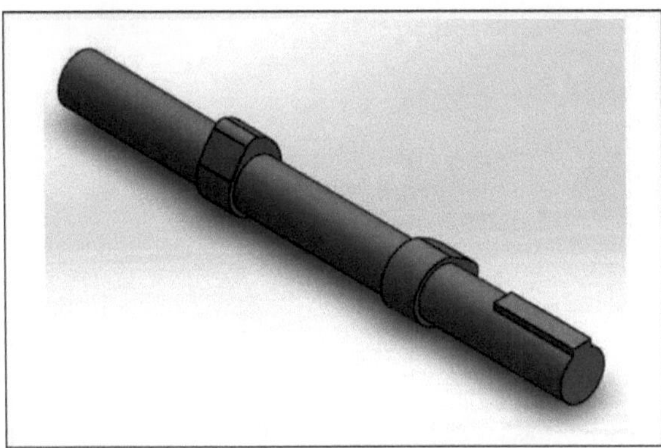

Source: Author.

Printed by Books on Demand GmbH, Norderstedt / Germany